AF700383

INTERNATIONAL CODE INTERPRETATIONS

2006

ICC
INTERNATIONAL CODE COUNCIL®

2006 International Code Interpretations®

ISBN # 1-58001-478-X

First Printing: September 2006

COPYRIGHT © 2006
by
INTERNATIONAL CODE COUNCIL, INC.

ALL RIGHTS RESERVED. This 2006 International Code Interpretations® is a copyrighted work owned by the International Code Council, Inc. Without advance written permission from the copyright owner, no part of this book may be reproduced, distributed or transmitted in any form or by any means, including, without limitation, electronic, optical or mechanical means (by way of example and not limitation, photocopying, or recording by or in an information storage retrieval system). For information on permission to copy material exceeding fair use, please contact: Publications, 4051 West Flossmoor Road, Country Club Hills, IL 60478-5771 (Phone 888-422-7233).

Trademarks: *"International Code Council,"* the *"International Code Council"* logo and the *"International Building Code," "International Residential Code," "International Fire Code," "International Mechanical Code," "International Plumbing Code," "International Existing Building Code," "International Energy Conservation Code,"* and the *"International Fuel Gas Code"* are trademarks of the *International Code Council, Inc.*

PRINTED IN THE U.S.A.

PREFACE

The 2006 *International Codes®* promulgated by the International Code Council® (ICC®) comprise a comprehensive set of model regulations for the construction and maintenance of safe buildings and structures. As a result of the performance nature of the *International Codes*, interpretations of the code provisions as they apply to a specific situation are sometimes necessary. Interpretation of the *International Codes* is a regular service offered to ICC members and is intended to be of an advisory nature, with final responsibility resting with the appropriate code official.

In the development of this document, a review of ICC or Legacy Code Committee Interpretations published to date was conducted. Those Interpretations that are no longer applicable because of changes in the code text since the 2000 editions of the ICC *International Codes* were omitted. Those in which the essential interpretative issue is still relevant to the 2006 editions were edited by ICC technical staff and revised as necessary for applicability.

The interpretations appearing in this book are in consecutive order according to code and section number. Applicable 2006 *International Code* text is included with each interpretation.

TABLE OF CONTENTS

International Building Code

International Residential Code

International Mechanical Code

International Code Interpretations

Building

2006

International Code Council

CHAPTER 1

ADMINISTRATION

SECTION 101
GENERAL

101.2 Scope. The provisions of this code shall apply to the construction, alteration, movement, enlargement, replacement, repair, equipment, use and occupancy, location, maintenance, removal and demolition of every building or structure or any appurtenances connected or attached to such buildings or structures.

> **Exception:** Detached one- and two-family dwellings and multiple single-family dwellings (townhouses) not more than three stories above grade plane in height with a separate means of egress and their accessory structures shall comply with the *International Residential Code*.

Q: Do code provisions apply to fences?

A: Yes. Section 101.2 states that the code shall apply to all buildings and structures. Fences are considered structures, and therefore are included within the scope of the code. The primary safety issue is structural integrity. Fences must be constructed to resist the anticipated live loads and installed in a manner ensuring structural stability. Chapter 33 also contains requirements for fences related to site work, demolition and construction. It should be noted that Section 105.2, Item 2 does exempt fences not over 6 feet high from needing a permit. It does not however exempt such fences from complying with the provisions of the *International Building Code*.

SECTION 104
DUTIES AND POWERS OF BUILDING OFFICIAL

104.7 Department records. The building official shall keep official records of applications received, permits and certificates issued, fees collected, reports of inspections, and notices and orders issued. Such records shall be retained in the official records for the period required for retention of public records.

Q: Do Sections 104.7 and 106.3.1 require construction documents to be kept on file with the code official even after construction has been completed?

A: No. Section 104.7 indicates only that the code official shall keep records of applications received, permits and certificates issued, fees collected, reports of inspections and notices and orders issued in the official records for as long as local or state laws require.

Although the code does not require construction documents to be retained, there may be state laws or other ordinances identifying information to be maintained in official records, which may include construction documents.

104.9 Approved materials and equipment. Materials, equipment and devices approved by the building official shall be constructed and installed in accordance with such approval.

Q: In the absence of either specific code provisions dealing with the installation of a product or the mention of the intended applicability in the permit application information, are the manufacturer's instructions and recommendations to be followed?

A: Yes, if the manufacturer's instructions and recommendations form the basis or a part of the basis for the code official's approval. Section 104.9 requires all materials, equipment and devices approved for use by the code official to be constructed and installed in accordance with that approval. The manufacturer's instructions and recommendations are to be followed if the material was approved for use with such consideration.

104.10 Modifications. Wherever there are practical difficulties involved in carrying out the provisions of this code, the building official shall have the authority to grant modifications for individual cases, upon application of the owner or owner's representative, provided the building official shall first find that special individual reason makes the strict letter of this code impractical and the modification is in compliance with the intent and purpose of this code and that such modification does not lessen health, accessibility, life and fire safety, or structural requirements. The details of action granting modifications shall be recorded and entered in the files of the department of building safety.

REFERENCED SECTION:

104.1 General. The building official is hereby authorized and directed to enforce the provisions of this code. The building official shall have the authority to render interpretations of this code and to adopt policies and procedures in order to clarify the application of its provisions. Such interpretations, policies and procedures shall be in compliance with the intent and purpose of this code. Such policies and procedures shall not have the effect of waiving requirements specifically provided for in this code.

Q: Is the authority to grant modifications as provided for in Section 104.10 intended to apply only to existing buildings?

A: No. Section 104.10 is included as part of the overall approval authority granted to the code official by Section 104.1 for use in dealing with the application of the *International Building Code*. The scope of the *International Building Code* includes both new and existing buildings.

SECTION 105
PERMITS

105.1 Required. Any owner or authorized agent who intends to construct, enlarge, alter, repair, move, demolish, or change the occupancy of a building or structure, or to erect, install, enlarge, alter, repair, remove, convert or replace any electrical, gas, mechanical or plumbing system, the installation of which is regulated by this code, or to cause any such work to be done, shall first make application to the building official and obtain the required permit.

Q: Is it the intent of Section 105.1 to require a building permit for the installation of new siding on an existing building?

A: Yes. The code regulates weather protection for buildings in Chapter 14 by setting forth requirements for materials and installation methods of exterior weather-resistant elements. Therefore, a permit is required to provide the code official with the opportunity to evaluate the installation of new siding for conformance to the applicable code requirements.

Q: If the owner of a proposed building or structure does not apply for a building permit, is the owner's registered design professional or contractor required, under Section 105.1, to obtain the permit?

A: Yes. A registered design professional or contractor empowered by the owner or lessee to act as the owner's agent would become one of the parties responsible for the permit application. However, in accordance with Section 105.1, a registered design professional or contractor must be specifically authorized by the owner to act as the owner's agent for taking out permits. If the owner does not empower an agent to act in such capacity, the responsibility remains with the owner.

Q: When specific code requirements that would be applicable to the work to be performed do not exist, does the adoption of the *International Building Code* provide the local jurisdiction with a mandate to require a building permit as stated in Section 105.1?

A: Yes. Section 105.1 applies to the work listed, regardless of the presence or lack of specific code requirements that are applicable to the specific project. There are exceptions where a permit is not required, but none of those exceptions relate to lack of specific applicable provisions. For instance, the demolition of a building is within the scope of the building code and is work requiring a permit as stated in Section 105.1, regardless of the fact that the specific provisions of Section 3303 might not apply to the project under review. This demolition work is still work which comes under the laws or ordinances of the adopting authority that has jurisdiction. The work requires review to ensure public health and safety. Items not specifically listed in the code could, indeed, be of concern to the jurisdiction, including such things as the presence of hazardous materials, such as asbestos, or some other health hazard, the sewer system, and the manner in which demolition operations are designed to ensure the safety of the crew.

105.4 Validity of permit. The issuance or granting of a permit shall not be construed to be a permit for, or an approval of, any violation of any of the provisions of this code or of any other ordinance of the jurisdiction. Permits presuming to give authority to violate or cancel the provisions of this code or other ordinances of the jurisdiction shall not be valid. The issuance of a permit based on construction documents and other data shall not prevent the building official from requiring the correction of errors in the construction documents and other data. The building official is also authorized to prevent occupancy or use of a structure where in violation of this code or of any other ordinances of this jurisdiction.

Q: Can a code official require compliance with the code, even though the approved construction documents contain information that is in conflict with the code, and a building permit has been issued?

A: Yes. Section 105.4 states that the permit shall not be construed to be a permit for, or an approval of, violation of any of the provisions of this code. Section 105.6 states that "the building official is authorized to suspend or revoke a permit issued under the provisions of this code wherever the permit is issued in error or on the basis of incorrect, inaccurate or incomplete information, or in violation of any ordinance or regulation or any of the provisions of this code." This points to the necessity for a comprehensive plan examination from the building department to minimize the difficulties this situation creates for both the building owner and code official.

Q: Does the code official have the authority to revisit a job after a final inspection has been made if a violation has been found and then take the necessary steps to have it corrected?

A: Yes. Section 105.4 indicates that a building permit is a license to build and shall not be construed as authority to violate, cancel or set aside any provisions of the code. It shall be unlawful to occupy a building contrary to or in violation of the code. Under no circumstances may the obligation to comply with the code be set aside.

105.5 Expiration. Every permit issued shall become invalid unless the work on the site authorized by such permit is commenced within 180 days after its issuance, or if the work authorized on the site by such permit is suspended or abandoned for a period of 180 days after the time the work is commenced. The building official is authorized to grant, in writing, one or more extensions of time, for periods not more than 180 days each. The extension shall be requested in writing and justifiable cause demonstrated.

Q: Is there a certain length of time that a building permit is valid after it has been issued?

A: As long as the authorized work is progressing, there was no intention for a limitation on the length of time that a permit is valid. Section 105.5 specifies that a permit becomes invalid if work is not commenced within 180 days after issuance or if the authorized work is suspended or abandoned for a period of 180 days after it begins.

SECTION 106 CONSTRUCTION DOCUMENTS

106.1 Submittal documents. Construction documents, statement of special inspections and other data shall be submitted in one or more sets with each application for permit. The construction documents shall be prepared by a registered design professional where required by the statutes of the jurisdiction in which the project is to be constructed. Where special conditions exist, the building official is authorized to require additional construction documents to be prepared by a registered design professional.

> **Exception:** The building official is authorized to waive the submission of construction documents and other data not required to be prepared by a registered design professional if it is found that the nature of the work applied for is such that review of construction documents is not necessary to obtain compliance with this code.

Q: Is it required under Sections 106.1 and 106.3.4.1 that plans for all building types, including single-family dwellings, bear the seal and signature of a registered design professional?

A: No. The necessity for a registered design professional's signature and seal on the plans would depend on whether special engineering details or computations were required. As stated in Section 106.1, it would also depend on registration and licensing laws of the state. Because many states employ different regulations, such as minimum building size, building use group, cost, etc., the professional registration laws must be reviewed to determine when a professional's seal and signature are required.

Q: Do the provisions of Section 106.1 require the design of a structural element, such as a floor or roof truss, to be performed only by a registered structural engineer?

A: No. Section 106.1 requires all engineering plans and computations to be signed by a registered design professional in accordance with the applicable state professional registration or licensing laws. The code does not specify the category of professional registration. The requirements of the state registration and licensing laws must be followed because they differ from state to state as to which category of registered design professional can perform various engineering functions. Some states do not have a registration category for structural engineers that is separate from other engineering disciplines.

106.3.1 Approval of construction documents. When the building official issues a permit, the construction documents shall be approved, in writing or by stamp, as "Reviewed for Code Compliance." One set of construction documents so reviewed shall be retained by the building official. The other set shall be returned to the applicant, shall be kept at the site of work and shall be open to inspection by the building official or a duly authorized representative.

Q: Do Sections 104.7 and 106.3.1 require construction documents to be kept on file with the code official even after construction has been completed?

A: No. Section 104.7 indicates only that the code official shall keep records of applications received, permits and certificates issued, fees collected, reports of inspections and notices and orders issued in the official records for as long as local or state laws require.

106.3.4.1 General. When it is required that documents be prepared by a registered design professional, the building official shall be authorized to require the owner to engage and designate on the building permit application a registered design professional who shall act as the registered design professional in responsible charge. If the circumstances require, the owner shall designate a substitute registered design professional in responsible charge who shall perform the duties required of the original registered design professional in responsible charge. The building official shall be notified in writing by the owner if the registered design professional in responsible charge is changed or is unable to continue to perform the duties.

The registered design professional in responsible charge shall be responsible for reviewing and coordinating submittal documents prepared by others, including phased and deferred submittal items, for compatibility with the design of the building.

Where structural observation is required by Section 1709, the statement of special inspections shall name the individual or firms who are to perform structural observation and describe the stages of construction at which structural observation is to occur (see also duties specified in Section 1704).

Q: Is it required under Sections 106.1 and 106.3.4.1 that plans for all building types, including single-family dwellings, bear the seal and signature of a registered design professional?

A: No. The necessity for a registered design professional's signature and seal on the plans would depend on whether special engineering details or computations were required. As stated in Section 106.1, it would also depend on registration and licensing laws of the state. Because many states employ different regulations, such as minimum building size, building use group, cost, etc., the professional registration laws must be reviewed to determine when a professional's seal and signature are required.

106.4 Amended construction documents. Work shall be installed in accordance with the approved construction documents, and any changes made during construction that are not in compliance with the approved construction documents shall be resubmitted for approval as an amended set of construction documents.

Q: Section 106.4 requires all work to conform to the approved construction documents, and any approved amendments, for which the permit was issued. Does this mean it is the code official's responsibility to ensure that all construction aspects conform to the approved plans, including those not related to health and safety and, therefore, not regulated by the code?

A: No. Section 105.3.1 stipulates, in part, that a permit is issued when the code official is satisfied that the proposed work conforms to code requirements. Many aspects of construction are not related to public health and safety and thus not regulated by the code, and are outside the domain of the code official. Section 101.3 further stipulates that the intent of the code is to establish the minimum requirements to safeguard the public health, safety and general welfare insofar as they are affected by building construction. Section 104.1 specifically states "The building official is hereby authorized and directed to enforce the provisions of this code." Beyond the code requirements, the code official has no expressed responsibility.

SECTION 112
BOARD OF APPEALS

112.2 Limitations on authority. An application for appeal shall be based on a claim that the true intent of this code or the rules legally adopted thereunder have been incorrectly interpreted, the provisions of this code do not fully apply, or an equally good or better form of construction is proposed. The board shall have no authority to waive requirements of this code.

REFERENCED SECTION:

104.10 Modifications. Wherever there are practical difficulties involved in carrying out the provisions of this code, the building official shall have the authority to grant modifications for individual cases, upon application of the owner or owner's representative, provided the building official shall first find that special individual reason makes the strict letter of this code impractical and the modification is in compliance with the intent and purpose of this code and that such modification does not lessen health, accessibility, life and fire safety, or structural requirements. The details of action granting modifications shall be recorded and entered in the files of the department of building safety.

Q: Does the board of appeals, as described in Section 112.2 of the code, have the authority to exempt or waive specific code provisions?

A: No. To appeal a decision of the code official, including when he or she does not grant a modification in accordance with Section 104.10, the owner of the building or structure, or any other person, may make an application to the board of appeals on the following basis:

1. The true intent of the code or the rules legally adopted thereunder have been incorrectly interpreted;
2. The code provisions do not fully apply; or
3. An equally good or better form of construction can be used.

It is not intended that the board of appeals set aside any code provisions without consideration of one of the above conditions.

CHAPTER 2

DEFINITIONS

SECTION 202 DEFINITIONS

BUILDING. Any structure occupied or intended for supporting or sheltering any use or occupancy.

Q: Would a structure consisting of a roof supported by columns and having no exterior walls be considered a building for code application purposes?

A: Yes. Based on the definitions of "Building," and "Area, building" in Sections 202 and 502.1, the above described structure is a building. Section 302.1 requires all buildings to be classified with respect to occupancy in one of the 10 listed groups. If the intended use is not specifically provided for in the code, the building or structure must be classified in the group that the occupancy use most closely resembles. The building must then comply with all code requirements based on that occupancy classification. The definition of "Area, building" specifically addresses how the area of a building without surrounding walls is determined for purposes of height and area limitations.

HABITABLE SPACE. A space in a building for living, sleeping, eating or cooking. Bathrooms, toilet rooms, closets, halls, storage or utility spaces and similar areas are not considered habitable spaces.

Q: Are prison and jail cells considered "habitable spaces" and therefore required to comply with the code provisions regarding habitable spaces?

A: No. The requirements for habitable spaces are intended to apply to residential occupancies and not to incarceration facilities. Although a prison or jail cell can accommodate occupancies generally found in a habitable space, both the functions and needs differ, and the application of the requirements can be counterproductive to the implied objectives. The construction of such facilities is usually deemed adequate when in compliance with the jurisdiction's requirements.

Q: When an existing basement or cellar area in a single-family dwelling unit is altered to create a "recreation room," do code requirements for "habitable spaces" apply?

A: Yes. Habitable spaces include areas used for living, sleeping, eating or cooking purposes. Recreation rooms can be characterized as being used for living purposes. If the area is used for sleeping purposes, the emergency escape and rescue requirements of Section 1026 of the *International Building Code* and Section R310 of the *International Residential Code* would also apply.

CHAPTER 3

USE AND OCCUPANCY

SECTION 303
ASSEMBLY GROUP A

303.1 Assembly Group A. Assembly Group A occupancy includes, among others, the use of a building or structure, or a portion thereof, for the gathering of persons for purposes such as civic, social or religious functions; recreation, food or drink consumption; or awaiting transportation.

Exceptions:

1. A building used for assembly purposes with an occupant load of less than 50 persons shall be classified as a Group B occupancy.
2. A room or space used for assembly purposes with an occupant load of less than 50 persons and accessory to another occupancy shall be classified as a Group B occupancy or as part of that occupancy.
3. A room or space used for assembly purposes that is less than 750 square feet (70 m^2) in area and is accessory to another occupancy shall be classified as a Group B occupancy or as part of that occupancy.

Assembly occupancies shall include the following:

A-1 Assembly uses, usually with fixed seating, intended for the production and viewing of the performing arts or motion pictures including, but not limited to:

Motion picture theaters
Symphony and concert halls
Television and radio studios admitting an audience
Theaters

A-2 Assembly uses intended for food and/or drink consumption including, but not limited to:

Banquet halls
Night clubs
Restaurants
Taverns and bars

A-3 Assembly uses intended for worship, recreation or amusement and other assembly uses not classified elsewhere in Group A including, but not limited to:

Amusement arcades
Art galleries
Bowling alleys
Places of religious worship
Community halls
Courtrooms
Dance halls (not including food or drink consumption)
Exhibition halls
Funeral parlors
Gymnasiums (without spectator seating)
Indoor swimming pools (without spectator seating)
Indoor tennis courts (without spectator seating)
Lecture halls
Libraries
Museums
Waiting areas in transportation terminals
Pool and billiard parlors

A-4 Assembly uses intended for viewing of indoor sporting events and activities with spectator seating including, but not limited to:

Arenas
Skating rinks
Swimming pools
Tennis courts

A-5 Assembly uses intended for participation in or viewing outdoor activities including, but not limited to:

Amusement park structures
Bleachers
Grandstands
Stadiums

Q: Is a building or facility used for showing, auctioning, and selling of nonhazardous merchandise classified as a Group A, Assembly Occupancy?

A: Yes. Any building or facility, or any room or space within a building or facility, used for the gathering of a prescribed number of people for amusement or recreation meets the definition for a Group A occupancy. The use of a room or facility where people assemble for the purpose of conducting an auction is classified as a Group A-3, Assembly Occupancy.

Q: What characteristics of an assembly occupancy serve to make Group A-2 nightclubs an appropriate occupancy classification?

A: The common factor in nightclub life losses has been high occupant densities. However, a large occupant load, by itself, does not provide the sole basis to distinguish between the level of hazard associated with a Group A-2 occupancy and any other assembly or educational occupancy.

The unique hazard level results from the presence of the following attributes of a nightclub-type occupancy:

1. Lighting levels
2. Entertainment type
3. Operation hours
4. Seating characteristics, including table sizes
5. Dance area
6. Service facilities

The low-lighting level, loud sound and associated activity, late hours, dense seating with ill-defined aisles and beverage service have all combined to delay fire recognition, confuse the appropriate response, increase egress time and contribute to extensive life losses.

SECTION 305
EDUCATIONAL GROUP E

305.2 Day care. The use of a building or structure, or portion thereof, for educational, supervision or personal care services for more than five children older than $2^1/_2$ years of age, shall be classified as a Group E occupancy.

Q: A church building contains a Sunday school classroom wing with two nursery rooms used only for a few hours at a time once a week.

Must these nursery rooms be considered an occupancy classification Group I-4 or E?

A: The occupancy classification is dependent on the age and number of individuals cared for in the nursery. If care is provided for more than five individuals who are more than $2^1/_2$ years old, the nursery is classified as Group E. If there are more than five children $2^1/_2$ years of age or younger, the nursery is classified as Group I-4 (see Section 308.5.2).

Q: A church building contains a Sunday school classroom wing with two nursery rooms used only for a few hours at a time once a week.

Would Section 508.3 regarding mixed uses and occupancies apply?

A: Yes. Because the occupancy classification of the nursery is either Group E or I-4 and the occupancy of the church is A-3, Section 508.3 provisions would apply.

Q: A church building contains a Sunday school classroom wing with two nursery rooms used only for a few hours at a time once a week.

Would the nursery room occupancy classification change if the rooms were occupied during the week as part of a day care facility?

A: No. Occupancy classification is not a function of the duration of a particular occupancy but rather of the hazard inherent to the occupants of the use itself.

SECTION 308
INSTITUTIONAL GROUP I

308.2 Group I-1. This occupancy shall include buildings, structures or parts thereof housing more than 16 persons, on a 24-hour basis, who because of age, mental disability or other reasons, live in a supervised residential environment that provides personal care services. The occupants are capable of responding to an emergency situation without physical assistance from staff. This group shall include, but not be limited to, the following:

Residential board and care facilities
Assisted living facilities
Halfway houses
Group homes
Congregate care facilities
Social rehabilitation facilities
Alcohol and drug centers
Convalescent facilities

A facility such as the above with five or fewer persons shall be classified as a Group R-3 or shall comply with the *International Residential Code* in accordance with Section 101.2. A facility such as above, housing at least six and not more than 16 persons, shall be classified as Group R-4.

Q: Is the Group I-1 classification appropriately applied to a building housing more than 16 occupants and resembling a multiple-family dwelling, but wherein:

- The building occupants are aged but not infirm; and
- Special emergency call switches monitored by health center staff are located in each dwelling unit?

A: No. Age, by itself, is not sufficient criterion for a Group I-1 occupancy classification. Group I-1 is only appropriate when occupants have a diminished mental capacity and must live in a supervised environment. The Group I-1 classification is intended to apply to conditions where occupants are physically capable of responding to an emergency situation without personal assistance but may not be mentally capable of an independent response. Emergency call switches in each dwelling unit are a convenience and do not necessarily indicate infirmity of the occupants.

Q: If a building classified and approved as a Group I-1 occupancy begins to house over five occupants with physical or mental ailments rendering them physically incapable of responding to an emergency situation without personal assistance, would this building have to be reclassified as Group I-2?

A: Yes. According to Section 308.2, a Group I-1 occupancy is intended to reflect the safety requirements appropriate for occupants that are physically capable of responding to an emergency situation without assistance.

A building with more than five occupants not possessing such capability represents a significantly different hazard, which is reflected in the code requirements that are applicable to Group I-2.

The introduction of more than five occupants, without the described response capability, into a building of a Group I-1 classification would represent a change in occupancy and would then be regulated in accordance with Section 3406.1. In both cases, the building would either have to be reclassified as Group I-2 or be restricted to individuals possessing the described response capability.

Q: If so, is Group I-2 the appropriate classification? The description of Group I-2 states physical, but not mental, limitations.

A: Yes. A Group I-2 occupancy classification would be appropriate because it is intended to include occupancy with six or more individuals having physical limitations. Group I-1 reflects the occupant's ability to respond physically to an emergency. Mentally unstable occupants who are unable to respond physically without assistance would be included in the scope of the Group I-2 description. Section 308.3 includes psychiatric care and mental hospitals within the scope of a Group I-2 classification.

308.3 Group I-2. This occupancy shall include buildings and structures used for medical, surgical, psychiatric, nursing or custodial care on a 24-hour basis for more than five persons who are not capable of self-preservation. This group shall include, but not be limited to, the following:

Hospitals
Nursing homes (both intermediate care facilities and skilled nursing facilities)
Mental hospitals
Detoxification facilities

A facility such as the above with five or fewer persons shall be classified as Group R-3 or shall comply with the *International Residential Code* in accordance with Section 101.2.

Q: If the facility provides 24-hour accommodations to ten persons, five of whom are not capable of self-preservation and five of whom are capable of responding to an emergency without assistance, is the facility classified as Group R-4?

A: No. A Group R-4 Residential Care/Assisted Living is a supervised residential facility that provides care for more than 5 occupants but not more than 16 occupants, all of whom must be "capable of responding to an emergency situation without physical assistance from staff." A facility that accommodates 5 or fewer such occupants is a Group R-3 occupancy, and a facility that accommodates more than 16 such occupants is a Group I-1 occupancy. A Group R-4 occupancy is not permitted to include any number of occupants that "are not capable of self-preservation."

A facility that accommodates more than five persons "who are not capable of self-preservation" is classified as a Group I-2 occupancy, whereas a facility that accommodates five or less occupants "who are not capable of self-preservation" is a Group R-3 occupancy.

A Group R-3 occupancy is permitted to accommodate a maximum of five occupants. The occupants may be "capable of responding to an emergency situation without physical assistance from staff" as permitted in a Group I-1 occupancy or "not capable of self-preservation" as required for a Group I-2 occupancy. A facility that accommodates 5 persons who are "capable of responding to an emergency situation without physical assistance from staff," and 5 persons "who are not capable of self-preservation" cannot be classified as a Group R-3 because the total occupant load of 10 persons exceeds the permitted maximum of 5 occupants.

The facility is a single occupancy; therefore, the entire facility must be assumed to be occupied by persons with the most restrictive capability when determining the occupancy classification of the facility. A facility that accommodates 10 occupants "who are not capable of self-preservation" is a Group I-2 occupancy.

Q: If an existing Group R-4 facility that provides 24-hour accommodations for 10 persons is expanded to accommodate an additional 6 persons who are not capable of responding to an emergency without physical assistance, must the occupancy classification of the existing facility be changed from a Group R-4 to Group I-2?

A: Yes. A facility that accommodates more than five occupants who are not capable of self preservation is classified as a Group I-2 occupancy. A Group R-4 facility is permitted to accommodate a maximum of 16 persons, but all of the occupants in a Group R-4 occupancy must be capable of self-preservation. The enlarged facility has an increased occupant load, which includes occupants who are not capable of self preservation; therefore, the entire occupant load must be assumed to be persons with the most restrictive capability.

The existing Group R-4 facility may be expanded and retain the Group R-4 occupancy classification, but the expanded portion of the existing facility that accommodates the six additional occupants who are not capable of self-preservation must comply with all the technical provisions for a Group I-2 occupancy. Because the existing portion of the facility is a Group R-4 occupancy and the expanded portion of the facility is a Group I-2 occupancy, the entire facility must comply with the provisions for mixed occupancies.

Q: Can a child care facility accommodating not more than five children 2 $^{1}/_{2}$ years old or younger and additional children over 2$^{1}/_{2}$ years old be classified as Group E, Educational occupancies?

A: Yes. Section 308.3.1 indicates that a facility accommodating *more* than five children 2 $^{1}/_{2}$ years old or younger is to be classified as Group I-2. A day care facility falling under the classification as described in Section 305.2 can include children 2 $^{1}/_{2}$ years old or younger and would not be reclassified as Group I-2 until the number of children exceeds five.

308.4 Group I-3. This occupancy shall include buildings and structures that are inhabited by more than five persons who are under restraint or security. An I-3 facility is occupied by persons who are generally incapable of self-preservation due to security measures not under the occupants' control. This group shall include, but not be limited to, the following:

Prisons
Jails
Reformatories
Detention centers
Correctional centers
Prerelease centers

Buildings of Group I-3 shall be classified as one of the occupancy conditions indicated in Sections 308.4.1 through 308.4.5 (see Section 408.1).

Q: Do Sections 308.4 and 408.2 require a room or building that is not used for sleeping purposes in a penal facility to be classified as a Group I-3 occupancy if some of the occupants are controlled for security reasons?

A: No. The five occupancy conditions listed in Sections 308.4.1 through 308.4.5 are specific to "sleeping areas," "activity spaces" and other "occupied areas within the smoke compartments." However, penal facilities are often a complex that does not necessarily have a Group I-3 classification in each building but still requires high levels of security throughout the structure. Section 408.2 specifically acknowledges that a mixed use condition may exist in penal facilities. The code requires that each portion of a building be classified according to its occupancy as long as the applicable provisions for mixed occupancies are met and the required means of egress is provided.

Section 1008.1.8.5, Exception 1 provides for maintaining inmate confinement at all other times. Buildings in a penal facility where the movements of occupants must be controlled for security reasons and that are primarily residential in usage must be classified as Group I-3. Other buildings within the complex must be classified according to their actual nonresidential uses, such as recreational, business, educational or industrial. Section 1008.1.8.5, Exception 1 only permits locked means of egress doors for security reasons for other buildings within a penal facility. Chapter 3 and Section 408.2 must be used to classify properly nonresidential buildings according to their occupancy versus whether the means of egress doors are locked in accordance with Section 1008.1.8.5, Exception 1.

SECTION 309 MERCANTILE GROUP M

309.1 Mercantile Group M. Mercantile Group M occupancy includes, among others, buildings and structures or a portion thereof, for the display and sale of merchandise, and involves stocks of goods, wares or merchandise incidental to such purposes and accessible to the public. Mercantile occupancies shall include, but not be limited to, the following:

Department stores
Drug stores
Markets
Motor fuel-dispensing facilities
Retail or wholesale stores
Sales rooms

Q: Is a Group M occupancy the appropriate classification for a building in which auction sales or other highly concentrated occupant load sales activities are conducted?

A: Yes. Section 309.1 describes mercantile buildings as those occupied for display and sales purposes involving stocks of goods, wares or merchandise that is incidental to such purposes, as well as accessible to the public. Although buildings housing auction sales do not sell goods by conventional retail methods, the merchandise is available for viewing prior to sale in areas accessible to the general public. Therefore, these types of buildings must comply with the criteria for a Group M occupancy. The presence of highly concentrated occupant loads does not, in itself, mandate an assembly use classification.

SECTION 312 UTILITY AND MISCELLANEOUS GROUP U

312.1 General. Buildings and structures of an accessory character and miscellaneous structures not classified in any specific occupancy shall be constructed, equipped and maintained to conform to the requirements of this code commensurate with the fire and life hazard incidental to their occupancy. Group U shall include, but not be limited to, the following:

Agricultural buildings
Aircraft hangars, accessory to a one- or two-family

residence (see Section 412.3)
Barns
Carports
Fences more than 6 feet (1829 mm) high
Grain silos, accessory to a residential occupancy
Greenhouses
Livestock shelters
Private garages
Retaining walls
Sheds
Stables
Tanks
Towers

Q: Are accessory occupancies classified as a Group U occupancy subject to the same snow load requirements as buildings of other occupancy classifications?

A: Yes. The code does not make a distinction between accessory and principal occupancies relative to structural load requirements for the building. Section 312.1 establishes that occupancies classified as utility and miscellaneous structures shall be constructed, equipped and maintained to conform to the code requirements that are commensurate with the fire and life hazards incidental to their occupancy. The structural design requirements for roofs are the minimum deemed necessary to withstand such elements. To allow construction of a building with an accessory occupancy that could reasonably be expected to collapse under the snow loads known to prevail in a fire area is not in the best interest of public safety.

CHAPTER 4

SPECIAL DETAILED REQUIREMENTS BASED ON USE AND OCCUPANCY

SECTION 402
COVERED MALL BUILDINGS

402.2 Definitions. The following words and terms shall, for the purposes of this chapter and as used elsewhere in this code, have the meanings shown herein.

ANCHOR BUILDING. An exterior perimeter building of a Group other than H having direct access to a covered mall building but having required means of egress independent of the mall.

COVERED MALL BUILDING. A single building enclosing a number of tenants and occupants such as retail stores, drinking and dining establishments, entertainment and amusement facilities, passenger transportation terminals, offices, and other similar uses wherein two or more tenants have a main entrance into one or more malls. For the purpose of this chapter, anchor buildings shall not be considered as a part of the covered mall building.

FOOD COURT. A public seating area located in the mall that serves adjacent food preparation tenant spaces.

GROSS LEASABLE AREA. The total floor area designed for tenant occupancy and exclusive use. The area of tenant occupancy is measured from the centerlines of joint partitions to the outside of the tenant walls. All tenant areas, including areas used for storage, shall be included in calculating gross leasable area.

MALL. A roofed or covered common pedestrian area within a covered mall building that serves as access for two or more tenants and not to exceed three levels that are open to each other.

402.5 Mall width. For the purpose of providing required egress, malls may be considered as corridors but need not comply with the requirements of Section 1005.1 of this code where the width of the mall is as specified in this section.

Q: Does Section 402.5 regulate the construction of fixed benches, trash containers and similar items in the mall area of a covered mall building?

A: No. Benches, trash containers and similar fixtures are considered furnishings and not part of the building or structure. Therefore, construction materials would not be regulated, provided they do not constitute an unusual fire hazard. However, Section 402.5.1 regulates the placement of furniture with regard to possible obstruction to the means of egress.

402.7.2 Tenant separation. Each tenant space shall be separated from other tenant spaces by a fire partition complying with Section 708. A tenant separation wall is not required between any tenant space and the mall.

Q: A kiosk is located in a covered mall with 10-foot clearance on each side. Is a 1-hour tenant separation required for the kiosk?

A: No. Separation is not required between a tenant space and mall. Therefore, the tenant separation is not required, since the kiosk is surrounded on all four sides by the mall.

Q: Is a large furniture display permitted in a covered mall, provided there is 10-foot clearance on each side?

A: Yes. The furniture display is no different than any of the other Mercantile occupancies located in the covered mall building. The display would be considered the same as a kiosk.

402.7.3 Anchor building separation. An anchor building shall be separated from the covered mall building by fire walls complying with Section 705.

> **Exception:** Anchor buildings of not more than three stories above grade plane that have an occupancy classification the same as that permitted for tenants of the covered mall building shall be separated by 2-hour fire-resistive fire barriers complying with Section 706.

402.7.3.1 Openings between anchor building and mall. Except for the separation between Group R-1 sleeping rooms and the mall, openings between anchor buildings of Types IA, IB, IIA and IIB construction and the mall need not be protected.

Q: Is a fire wall required in a covered mall building between a tenant space and an anchor store?

A: Yes. Section 402.7.3 specifically requires a fire wall. However, the exception allows a fire-resistive reduction for under certain conditions to a 2-hour fire-resistive fire barrier.

402.10 Kiosks. Kiosks and similar structures (temporary or permanent) shall meet the following requirements:

1. Combustible kiosks or other structures shall not be located within the mall unless constructed of any of the following materials:
   1.1. Fire-retardant-treated wood complying with Section 2303.2.
   1.2. Foam plastics having a maximum heat release rate not greater than 100kW (105 Btu/h) when tested in accordance with the exhibit booth protocol in UL 1975.
   1.3. Aluminum composite material (ACM) having a flame spread index of not more than 25 and a smoke-developed index of not more than 450 when tested as an assembly in the maximum thickness intended for use in accordance with ASTM E 84.
2. Kiosks or similar structures located within the mall shall be provided with approved fire suppression and detection devices.
3. The minimum horizontal separation between kiosks or groupings thereof and other structures within the mall shall be 20 feet (6096 mm).
4. Each kiosk or similar structure or groupings thereof shall have a maximum area of 300 square feet (28 m^2).

Q: Are store fixtures that are located in the mall area of a covered mall building with no superstructure around them considered kiosks?

A: Yes. The intent of Section 402.10 is to regulate the distribution of combustibles within the mall. The fuel load is traditionally located in a kiosk. "Kiosk" is defined in *Webster's Dictionary* as, "a small, light structure with one or more open sides." When a substantial fuel load is located within a mall without being enclosed in a kiosk, it is still necessary to limit the maximum area of the fuel load to 300 square feet and the horizontal separation between arrangements of combustibles to 20 feet. While the code does not require that a kiosk be enclosed, it is appropriate to apply the same limitations as specified in the code for kiosks to such an arrangement.

Q: Are temporary displays, such as a "Santa house," a stage for choirs, a model home for a house and flower show and other similar displays, required to comply with Section 402.10?

A: Yes. Whether for retail, decorative or other use, it is appropriate for groups of combustible materials to be regulated in a consistent manner and planned for in the design of the mall area.

SECTION 403
HIGH-RISE BUILDINGS

403.1 Applicability. The provisions of this section shall apply to buildings having occupied floors located more than 75 feet (22 860 mm) above the lowest level of fire department vehicle access.

Exception: The provisions of this section shall not apply to the following buildings and structures:

1. Airport traffic control towers in accordance with Section 412.
2. Open parking garages in accordance with Section 406.3.
3. Buildings with an occupancy in Group A-5 in accordance with Section 303.1.
4. Low-hazard special industrial occupancies in accordance with Section 503.1.1.
5. Buildings with an occupancy in Group H-1, H-2 or H-3 in accordance with Section 415.

Q: When a building consists of seven stories, each 12 feet in height, for a total building height of 84 feet, each story being habitable or occupiable, and grade is level at all elevations, must this building conform to the high-rise building provisions of Section 403?

A: No. The applicability criteria for the high-rise building provisions of Section 403 are based on buildings having occupied floors located more than 75 feet above the lowest level of vehicle access for fire departments in accordance with Section 403.1. The topmost occupiable floor level of the building described is 72 feet above the level of vehicle access and, therefore, the building is not subject to high-rise provisions.

Q: What or where is the lowest reference point used to determine the level that is 75 feet above the lowest level of vehicle access for fire departments?

A: The intent of Section 403.1 is to base the application of high-rise building provisions on the lowest level at which a fire department vehicle could be staged at the exterior for carrying out fire-fighting operations. It is not intended to be measured as 75 feet from the highest point of an aerial ladder, but rather taken from the lowest ground level on which the vehicle would be located.

403.3 Reduction in fire-resistance rating. The fire-resistance-rating reductions listed in Sections 403.3.1 and 403.3.2

shall be allowed in buildings that have sprinkler control valves equipped with supervisory initiating devices and water-flow initiating devices for each floor.

Q: If sprinklers are omitted from bathrooms and closets in a high-rise building with an occupancy in Group R-1, do the automatic sprinkler system alternatives of Section 403.3 still apply?

A: Yes. Whether or not sprinkler alternatives are utilized by the designer and owner, Sections 403.2 and 403.3 require the system to be designed in accordance with Section 5.13.9 (NFPA 13), which permits the omission of sprinklers from guestroom bathrooms not exceeding 55 square feet in area and from guestroom closets not exceeding 24 square feet in area.

SECTION 404 ATRIUMS

404.1 General. In other than Group H occupancies, and where permitted by Exception 5 in Section 707.2, the provisions of this section shall apply to buildings or structures containing vertical openings defined herein as "Atriums."

Q: Section 5-13.4 of NFPA 13 specifies requirements for draftstops and sprinklers around unenclosed floor openings. Is this section applicable to an atrium or other unenclosed floor openings for conformance to the requirements of Sections 404.1 and 707.2?

A: No. Section 102.4 specifies that when differences occur between specific code provisions and a referenced standard, code provisions shall apply. Sections 404 and 707.2 clearly establish mandatory requirements for the protection of either atriums or other unenclosed floor openings. Therefore, specific code provisions take precedence over Section 5-13.4 of NFPA 13.

404.5 Enclosure of atriums. Atrium spaces shall be separated from adjacent spaces by a 1-hour fire barrier constructed in accordance with Section 706 or a horizontal assembly constructed in accordance with Section 711, or both.

Exceptions:

1. A glass wall forming a smoke partition where automatic sprinklers are spaced 6 feet (1829 mm) or less along both sides of the separation wall, or on the room side only if there is not a walkway on the atrium side, and between 4 inches and 12 inches (102 mm and 305 mm) away from the glass and designed so that the entire surface of the glass is wet upon activation of the sprinkler system without obstruction. The glass shall be installed in a gasketed frame so that the framing system deflects without breaking (loading) the glass before the sprinkler system operates.
2. A glass-block wall assembly in accordance with Section 2110 and having a $^{3}/_{4}$-hour fire protection rating.
3. The adjacent spaces of any three floors of the atrium shall not be required to be separated from the atrium where such spaces are included in the design of the smoke control system.

Q: Is it the intent of Exception 1 to Section 404.5 to require a deluge sprinkler system to protect glass walls separating an atrium from adjacent spaces?

A: No. The specially designed sprinkler system located at the glass wall is intended to protect the glazing material from breakage as a result of thermal shock. It is not necessary to activate all of the sprinklers along the glazing material to provide such protection so long as the entire surface of the glazed panel is wetted by the sprinkler system.

Q: What design criteria are required for such sprinklers?

A: NFPA 13 provides two design options: pipe schedule sizing and hydraulically designed systems. Hydraulically designed systems should be evaluated to determine if the most hydraulically remote area occurs adjacent to the glazing as a result of the closer spacing of sprinklers at this wall. In accordance with Section 7-2.2.1 of NFPA 13, the use of pipe schedule designed systems is limited to new installations of 5,000 square feet or less or for additions or modifications to existing pipe schedule systems.

SECTION 414 HAZARDOUS MATERIALS

414.2.4 Fire-resistance-rating requirements. The required fire-resistance rating for fire barriers shall be in accordance with Table 414.2.2. The floor construction of the control area and the construction supporting the floor of the control area shall have a minimum 2-hour fire-resistance rating.

Exception: The floor construction of the control area and the construction supporting the floor of the control area is allowed to be 1-hour fire-resistance rated in buildings of Type IIA, IIIA and VA construction, provided that both of the following conditions exist:

1. The building is equipped throughout with an automatic sprinkler system in accordance with Section 903.3.1.1, and
2. The building is three stories or less in height.

Q: A control area is located within a building that is permitted to be constructed using building elements with a required fire-resistance rating equal to or less than that required by Table 414.2.2. Is the minimum 2-hour fire-resistance-rated construction required in Section 414.2.2 required to include the entire floor structure of the floor on which the control area is located as well as all construction supporting the floor where the control area is located?

A: No. Only the building elements within the load path of the portion of the floor where the control area is located are subject to the provisions of Section 414.2.2. It is not necessary to provide 2-hour fire-resistance-rated construction for the entire floor and supporting structure where the control area is located. The purpose for allowing 2-hour fire-resistance-rated construction to be isolated to the specific building elements providing structural support for a control area permits control areas to be located in buildings with unprotected types of construction, such as IIB, IIIB, and VB, without requiring a change of construction type for the entire building.

[F] 414.5.1 Explosion control. Explosion control shall be provided in accordance with the *International Fire Code* as required by Table 414.5.1 where quantities of hazardous materials specified in that table exceed the maximum allowable quantities in Table 307.1(1) or where a structure, room or space is occupied for purposes involving explosion hazards as required by Section 415 or the *International Fire Code*.

Q: Is it the intent of Section 414.5.1 to require the installation of explosion venting and explosion suppression in occupancies containing hazardous materials where flammable liquids are transferred during a container filling operation?

A: No. Section 414.5.1 requires explosion venting and explosion suppression for structures or spaces occupied for purposes "involving explosion hazards." When flammable liquid vapors are present in concentration over one-fourth of their lower flammable limit, an explosion hazard exists. NFPA 30, the referenced standard for flammable and combustible liquids, requires ventilation systems and equipment to limit flammable vapor-air mixtures to prevent a possible explosion hazard. Thus, the presence of an explosion hazard would not exist in a code-complying installation, and the requirements of Section 414.5.1 would not apply.

SECTION 415
GROUPS H-1, H-2, H-3, H-4 AND H-5

[F] 415.6.1.2 Grinding rooms. Every room or space occupied for grinding or other operations that produce combustible dusts shall be enclosed with fire barriers that have not less than a 2-hour fire-resistance rating where the area is not more than 3,000 square feet (279 m^2), and not less than a 4-hour fire-resistance rating where the area is greater than 3,000 square feet (279 m^2).

Q: Section 415.6.1.2 requires all grinding rooms exceeding 3,000 square feet in area to have enclosing floors and walls of not less than a 4-hour fire-resistance rating. Are the exterior walls of a grinding room required to have a fire-resistance rating of 4 hours?

A: No. The intent of requiring 4-hour fire-resistance-rated floors and walls is to provide separation of the grinding room from other interior spaces of the building. The exterior walls must comply with the requirements of Table 602 and Section 704.8 for the type of construction, load-bearing capability, occupancy classification and fire separation distance.

CHAPTER 5

GENERAL BUILDING HEIGHTS AND AREAS

SECTION 502 DEFINITIONS

502.1 Definitions. The following words and terms shall, for the purposes of this chapter and as used elsewhere in this code, have the meanings shown herein.

AREA, BUILDING. The area included within surrounding exterior walls (or exterior walls and fire walls) exclusive of vent shafts and courts. Areas of the building not provided with surrounding walls shall be included in the building area if such areas are included within the horizontal projection of the roof or floor above.

Q: Is the definition of "Area, building" in Section 502.1 intended to include the thicknesses of a building's exterior walls to determine compliance with height and area limitations?

A: No. Building area is intended to include only such area within or between surrounding exterior walls or exterior walls and fire/party walls. However, any area contained within the maximum horizontally projected area due to the story above is to be included. Where the roof or other architectural projections extend past the exterior wall, the area below shall also be considered a building area if the space is classified in one of the occupancies listed in Section 302.1. Height and area limitations are a function of both the occupancy hazard and fire-resistance-rated construction contained within the building rather than in the exterior shell itself.

HEIGHT, BUILDING. The vertical distance from grade plane to the average height of the highest roof surface.

Q: In determining the height of a building with a gabled roof, is it permissible to use the centroid of the projected triangular shape of the roof as the "average height" of Section 502.1?

A: No. Although the centroid of the projected triangular shape would result in a less-restrictive building height, the ordinarily accepted meaning of "average height" would be the distance midway between the extremes. For a gable roof configuration, the building height would be the distance measured from the grade plane to one-half the distance between the eave and ridge of the gable end.

MEZZANINE. An intermediate level or levels between the floor and ceiling of any story and in accordance with Section 505.

Q: Assume a building is equipped with storage racks that have a 36-inch-wide walkway located 8 feet above the ground between racks. The racks and shelves are used for storage, and the walkway provides access to the upper shelves. The racks, which are not considered a structural part of the building, are erected by the owner after occupancy.

Would this walkway be considered a "mezzanine" as defined in Section 502.1?

A: No. Such a walkway is not considered a mezzanine, because the shelves and racks do not represent a structural element of the building and the use of the walkway is limited to servicing the building contents.

SECTION 503 GENERAL HEIGHT AND AREA LIMITATIONS

503.1.2 Buildings on same lot. Two or more buildings on the same lot shall be regulated as separate buildings or shall be considered as portions of one building if the height of each building and the aggregate area of buildings are within the limitations of Table 503 as modified by Sections 504 and 506. The provisions of this code applicable to the aggregate building shall be applicable to each building.

Q: When two buildings on the same lot are considered portions of one aggregate building in accordance with Section 503.1.2, are the opposing enclosure walls considered exterior walls for fire-resistance-rating purposes?

A: No. When evaluated as a single aggregate building, two enclosure walls facing each other are considered equivalent to a single interior wall, and the required fire-resistance rating should be based on the appropriate requirements of Table 601 regarding interior walls. However, such walls still must comply with the exterior envelope requirements of Chapter 14, providing each building with the required weather protection.

SECTION 505
MEZZANINES

505.2 Area limitation. The aggregate area of a mezzanine or mezzanines within a room shall not exceed one-third of the floor area of that room or space in which they are located. The enclosed portion of a room shall not be included in a determination of the floor area of the room in which the mezzanine is located. In determining the allowable mezzanine area, the area of the mezzanine shall not be included in the floor area of the room.

Exceptions:

1. The aggregate area of mezzanines in buildings and structures of Type I or II construction for special industrial occupancies in accordance with Section 503.1.1 shall not exceed two-thirds of the area of the room.
2. The aggregate area of mezzanines in buildings and structures of Type I or II construction shall not exceed one-half of the area of the room in buildings and structures equipped throughout with an approved automatic sprinkler system in accordance with Section 903.3.1.1 and an approved emergency voice/alarm communication system in accordance with Section 907.2.12.2.

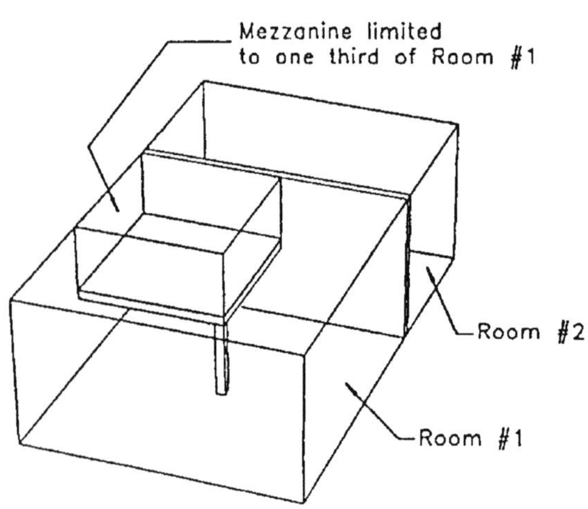

Figure 505.2

Q: A wall that encloses a mezzanine in accordance with Exception 2 of Section 505.4 completely subdivides the building as shown in Figure 505.2. Insofar as Exception 2 does not require the mezzanine to be completely open to the room in which it is located, can the allowable size of the mezzanine be determined based on the aggregate floor area of the building?

A: No. In this situation, the building is subdivided into two distinct rooms. Section 505.2 limits the allowable size of the mezzanine to one-third of the area of the one room in which it is located. The sole intent of Exception 2 of Section 505.4 is to define the conditions by which a mezzanine can be completely enclosed. This exception is not intended to allow a larger mezzanine than permitted by Section 505.2.

Also note that if the space below the mezzanine is also enclosed, that area must be excluded from the room area used to determine the mezzanine's maximum size (see Section 505.2).

Q: Is the maximum allowable area of a mezzanine one-third of the floor area of the open space of the story below the mezzanine?

A: Yes. If you have a one-story building of 5,000 square feet and an open area inside this building of 3,000 square feet, the maximum allowable area of the mezzanine would be one-third of 3,000 square feet, which is 1,000 square feet. The aggregate area of all mezzanines in this 3,000 square feet cannot exceed 1,000 square feet without the mezzanine constituting the second story.

505.3 Egress. Each occupant of a mezzanine shall have access to at least two independent means of egress where the common path of egress travel exceeds the limitations of Section 1014.3. Where a stairway provides a means of exit access from a mezzanine, the maximum travel distance includes the distance traveled on the stairway measured in the plane of the tread nosing. Accessible means of egress shall be provided in accordance with Section 1007.

Exception: A single means of egress shall be permitted in accordance with Section 1015.1.

Q: Does Section 1020.1 require stairs from a mezzanine level to the room below to be enclosed in fire separation assemblies?

A: No. Mezzanines in compliance with Section 505 are considered to be a portion of the floor below. Therefore, the stairway from the mezzanine serves the same purpose as an exit access doorway from any other room within that floor area. As long as the exit access travel distance from the mezzanine to an exit serving the floor area below it does not exceed the distances referenced in Table 1016.1, the stairway from the mezzanine is part of the exit access. Only required interior exit stairways must be enclosed in fire separation assemblies complying with Section 1020.1 (see Exception 2 to Section 505.4).

Q: Does Section 1024.1 require stairs from a mezzanine level to adjoin a public way, level of exit discharge passageway or other unobstructed exit element?

A: No. A mezzanine stair treated as an exit access element is not required to conform to Section 1024.1, which only applies to exit elements discharging within the building.

SECTION 506
AREA MODIFICATIONS

506.1 General. The areas limited by Table 503 shall be permitted to be increased due to frontage (If) and automatic sprinkler system protection (Is) in accordance with the following:

$$A_a = \left\{A_t + \left[A_t'' \ I_f\right] + \left[A_t'' \ I_s\right]\right\} \qquad \textbf{(Equation 5-1)}$$

where:

A_a = Allowable area per story (square feet).

A_t = Tabular area per story in accordance with Table 503 (square feet).

I_f = Area increase factor due to frontage as calculated in accordance with Section 506.2.

I_s = Area increase factor due to sprinkler protection as calculated in accordance with Section 506.3.

Q: In determining the allowable building area when several area modifications are applicable, are the modifications applied cumulatively?

A: No. The area increases are applied as a total percentage multiplier of the tabular area as specified in Table 503. For example, if a 75-percent frontage increase and a 200-percent automatic sprinkler system increase are applicable, the allowable building area computation would be as follows:

Allowable building area = Tabular area +
[(75% + 200%) Tabular area]

506.2 Frontage increase. Every building shall adjoin or have access to a public way to receive an area increase for frontage. Where a building has more than 25 percent of its perimeter on a public way or open space having a minimum width of 20 feet (6096 mm), the frontage increase shall be determined in accordance with the following:

$$I_f = \left[F / P - 0.25\right] W / 30 \qquad \textbf{(Equation 5-2)}$$

where:

I_f = Area increase due to frontage.

F = Building perimeter that fronts on a public way or open space having 20 feet (6096 mm) open minimum width (feet).

P = Perimeter of entire building (feet).

W = Width of public way or open space (feet) in accordance with Section 506.2.1.

Q: Is a fire wall that subdivides one building into two separate structures considered "fronting on a street or other unoccupied space" for the purpose of computing the street frontage increase in accordance with Section 506.2?

A: No. The 20-foot-minimum unoccupied space used for street frontage area increases must be "unoccupied" and have "access" from a street primarily for fire-fighting purposes. A fire wall does not perform this function and thus cannot be considered an open perimeter.

Q: When constructing an addition to an existing building, can the existing building perimeter be used, including that of the proposed addition to determine the street frontage increase in accordance with Section 506.2?

A: Yes. Section 3403.1 requires that when a building is increased in size, the existing portions, as well as a new addition, must not exceed the height and area limitations specified for new buildings. Therefore, the entire building perimeter (old and new) must be considered when calculating the area increase for a given type of construction. However, the portion of the existing building perimeter that is common to the building addition cannot be credited as being an unoccupied space with access from a street.

506.4 Area determination. The maximum area of a building with more than one story above grade plane shall be determined by multiplying the allowable area of the first story (A_a), as determined in Section 506.1, by the number of stories above grade plane as listed below:

1. For buildings with two stories above grade plane, multiply by 2;
2. For buildings with three or more stories above grade plane, multiply by 3; and
3. No story shall exceed the allowable area per story (A_a), as determined in Section 506.1, for the occupancies on that story.

Exceptions:

1. Unlimited area buildings in accordance with Section 507.
2. The maximum area of a building equipped throughout with an automatic sprinkler system in accordance with Section 903.3.1.2 shall be determined by multiplying the allowable area per story (A_a), as determined in Section 506.1, by the number of stories above grade plane.

Q: Assume a multistory building that has a different allowable area per floor (A_a) because the occupancy classification of each floor is different. The different occupancy classifications are separated with fire-resistive barriers/assemblies.

Is the maximum area of the building limited by adding the per story building ratios together, dividing by the number of stories (up to a maximum of three stories) and the resulting ratio cannot exceed a value of 1?

A: Yes. However, the area ratio for each floor, as stipulated in Section 508.3.3, shall not exceed a value of 1. The area ratio for each floor is determined by applying the unity formula where the Actual Area is divided by the Allowable Area (Table 503). The resulting value is the area ratio for that floor, which shall not exceed 1.

Section 506.4 limits the maximum area of a building to the allowable area per floor multiplied by a maximum of three floors. For a multistory, single occupancy building, the area ratio is not applied. The area ratio is applied to multistory, mixed occupancy buildings, and the sum of the area ratios of all floors divided by 3 shall not exceed 1; or more simply stated, the sum of the area ratios of all floors shall not exceed 3.

The area ratio (unity formula) does not apply to buildings designed using the nonseparated provisions of Section 508.3.2.1.

Understanding the Unity Formula (Mixed-Area Ratio) Section 506.4, *International Building Code*

In applying the provisions of the *International Building Code,* one must understand that many of the separate provisions must be used together. One example is the "unity formula" , so termed because it deals with the allowable area ratio of mixed occupancy buildings in which one or more fire barriers separate the individual occupancies. Under the *International Building Code*, an architect may choose whether to or not to separate areas of a building that house different uses or occupancies. The exception is that Group H occupancies must be in mixed-use buildings, must be separated by appropriate fire barriers and may not be in unseparated buildings. (Note that Group H-1 occupancies must be in single-use buildings and may not be in mixed-use buildings).

Section 508.3.3.2 of the *International Building Code* states, "In each story, the building area shall be such that the sum of the ratios of the actual floor area of each occupancy divided by the allowable area of each occupancy shall not exceed 1." Thus, the unity formula impacts the allowable area of any given floor or story of a building. The unity formula may be expressed as follows:

$$\frac{\text{Actual Area}}{\text{Allowable Area}} + \frac{\text{Actual Area}}{\text{Allowable Area}} + \frac{\text{Actual Area}}{\text{Allowable Area}} \leq 1$$

The diagram below depicts a small strip center of Type IIB construction with use groups as indicated. Assuming that there are no increases in the allowable area, use the unity formula to check for compliance.

Group M	Group B	Group A-3	Group E
Retail	Office	Library	Day Care
3,750 ft.^{2}	4,600 ft.^{2}	3,230 ft.^{2}	2,320 ft.^{2}

Using *International Building Code* Table 503, the allowable area of each use-group is as follows:

Group M: 12,500 ft.^{2} Group B: 23,000 ft.^{2}

Group A-3: 9,500 ft.^{2} Group E: 14,500 ft.^{2}

Apply the unity formula:

$$\frac{3{,}750}{12{,}500} + \frac{4{,}600}{23{,}000} + \frac{3{,}230}{9{,}500} + \frac{2{,}320}{14{,}500} \leq ?$$

$$0.3 + 0.2 + 0.34 + 0.16 \leq 1$$

The resulting value is 1, so the building under consideration complies with the code with regard to allowable area. Provided that there were abilities to increase the area based on open yards or sprinklers, the maximum allowable area for each separate occupancy must be determined using Section 506 of the *International Building Code*.

With multistory occupancies, the unity formula must be applied to each separate story, with the result being a value less than or equal to 1. Buildings that exceed three stories, however, require additional consideration. *International Building Code* Section 506.4 limits the maximum allowable area of a building to three times that determined for a single-story building. Thus, the area ratio calculation for each floor must be less than or equal to 1, and the ratio area calculation for the total area of all floors may not exceed 3. In the example below, assume the building and each occupancy receive the maximum increase permitted for frontage increase, 75 percent (per Section 506.2), and a fire sprinkler increase of 200 percent (per Section 506.3). The building is of Type IIA construction.

The allowable area per floor per occupancy based on Equation 5-1, Section 506.1, is as follows:

Group A-3: 58,125 ft.^{2}; Group R-2: 90,000 ft.^{2}; Group A-2: 58,125 ft.^{2};

Group R-1: 90,000 ft.^{2}; Group R-2:- 90,000 ft.^{2};

Group B : 140,625 ft.^{2}; Group M: 80,625 ft.^{2}; Group R-1: 90,000 ft.^{2};

Group B - 140,625 ft.^{2}; Group M: 80,625 ft.^{2}

For this four-story building, the value of the mixed-area ratio for each floor does not exceed 1, and the total value for all the floors in the building does not exceed 3. Therefore, the building is within allowable area.

The unity formula does not apply to buildings designed using the nonseparated provisions of the *International Building Code* Section 508.3.2.

SECTION 507
UNLIMITED AREA BUILDINGS

507.1 General. The area of buildings of the occupancies and configurations specified herein shall not be limited.

This interpretation pertains to all sections addressed in Section 507 Unlimited Area Buildings. Text of each individual section listed below is not included.

507.2 Nonsprinklered, one story.

507.3 Sprinklered, one story.

507.4 Two story.

507.5 Reduced open space.

507.6 Group A-3 buildings.

507.7 Group H occupancies.

507.8 Aircraft paint hangar.

507.9 Group E buildings.

507.10 Motion picture theaters.

Q: Are structures such as tanks, cooling towers, generators, and trash dumpsters permitted within the 60-foot public way or yard that is required to surround and adjoin an unlimited area building?

A: Yes. An unlimited area building must be surrounded by not less than 60 feet. The fire separation distance, as defined in Section 702.1, is the distance measured from the face of the unlimited area building to an interior lot line, the centerline of an adjacent street, alley or public way, or to an imaginary line between another building on the same property. No other building is permitted within the 60-foot fire separation distance. The purpose of the fire separation distance is to establish a minimum separation between buildings that will serve to reduce the possibility that a fire in one building can spread to another building. It is not the intent to prohibit elements other than buildings from being located within the fire separation distance.

Q: Is an automotive parking lot permitted within the 60-foot public way or yard that is required to surround and adjoin an unlimited area building?

A: Yes. A public way can contain a street or alley that is specifically dedicated for vehicular circulation and parking. Vehicular circulation and parking are permitted within the portion of the 60-foot fire separation distance that extends beyond the building property and into a public way. Likewise, vehicular circulation and parking are permitted within the portion of the 60-foot fire separation distance that is located within the building property.

Q: Is the 60-foot yard that is required to surround an unlimited area building required to be publicly owned?

A: No. The 60-foot fire separation (open space) must be located on the same property as the building or on property dedicated for public use. The property on which the building is located may be privately owned or publicly owned.

507.2 Nonsprinklered, one story. The area of a one-story, Group F-2 or S-2 building shall not be limited when the building is surrounded and adjoined by public ways or yards not less than 60 feet (18 288 mm) in width.

507.3 Sprinklered, one story. The area of a one-story, Group B, F, M or S building or a one-story Group A-4 building, of other than Type V construction, shall not be limited when the building is provided with an automatic sprinkler system throughout in accordance with Section 903.3.1.1 and is surrounded and adjoined by public ways or yards not less than 60 feet (18 288 mm) in width.

Exceptions:

1. Buildings and structures of Type I and II construction for rack storage facilities that do not have access by the public shall not be limited in height, provided that such buildings conform to the requirements of Sections 507.2 and 903.3.1.1 and NFPA 230.
2. The automatic sprinkler system shall not be required in areas occupied for indoor participant sports, such as tennis, skating, swimming and equestrian activities in occupancies in Group A-4, provided that:
    2.1. Exit doors directly to the outside are provided for occupants of the participant sports areas; and
    2.2. The building is equipped with a fire alarm system with manual fire alarm boxes installed in accordance with Section 907.
3. Group A-1 and A-2 occupancies of other than Type V construction shall be permitted, provided:
    3.1. All assembly occupancies are separated from other spaces as required for separated uses in Section 508.3.3.4 with no reduction allowed

in the fire-resistance rating of the separation based upon the installation of an automatic sprinkler system;

3.2. Each Group A occupancy shall not exceed the maximum allowable area permitted in Section 503.1; and

3.3. All required exits shall discharge directly to the exterior.

Q: Is it the intent of Sections 507.2 and 507.3 to apply to one-story buildings without basements or cellars?

A: Yes. Sections 507.2 and 507.3 provide an exception to the height and area provisions of Table 503 relative to one-story buildings. The primary consideration is that the complete perimeter would be open and available for fire department access and operations.

SECTION 508 MIXED USE AND OCCUPANCY

508.2 Incidental uses. Incidental use areas shall comply with the provisions of this section.

Exception: Incidental use areas within and serving a dwelling unit are not required to comply with this section.

Q: Is an "Incidental Use Area" identified as Table 508.2 permitted to be classified as an occupancy use group different from the building in which it is located, and the building classified as a mixed use building subject to the requirements of Section 508.3, Mixed occupancies?

A: Yes. Section 508.2.1 is intended to apply to specific areas that are considered to be part of the main occupancy classification, which generally represent a higher relative fire hazard. Section 508.2.1 is allowing these to be considered as part of the main occupancy, thus not requiring consideration of designing the building as if it contained a occupancy classification such as Group S or F, because the presence of this area with higher relative hazard is accounted for by requiring that area to be separated and/or protected by a fire suppression system.

However, when such areas are classified separately as to their occupancy, and the building is considered to be of mixed occupancy, the presence of the higher relative fire hazard is accounted for in a different way - by fire protection features that result from considering such areas as one of the building's multiple occupancy classifications. The application of Section 508.3 will result in certain building features being governed by code requirements commensurate with higher relative fire hazard or a storage or workshop area.

TABLE 508.2
INCIDENTAL USE AREAS

ROOM OR AREA	SEPARATION[a]
Furnace room where any piece of equipment is over 400,000 Btu per hour input	1 hour or provide automatic fire-extinguishing system
Rooms with any boiler over 15 psi and 10 horsepower	1 hour or provide automatic fire-extinguishing system
Refrigerant machinery rooms	1 hour or provide automatic sprinkler system
Parking garage (Section 406.2)	2 hours; or 1 hour and provide automatic fire-extinguishing system
Hydrogen cut-off rooms	1-hour fire barriers and floor/ceiling assemblies in Group B, F, H, M, S and U occupancies. 2-hour fire barriers and floor/ceiling assemblies in Group A, E, I and R occupancies.
Incinerator rooms	2 hours and automatic sprinkler system
Paint shops, not classified as Group H, located in occupancies other than Group F	2 hours; or 1 hour and provide automatic fire-extinguishing system
Laboratories and vocational shops, not classified as Group H, located in Group E or I-2 occupancies	1 hour or provide automatic fire-extinguishing system
Laundry rooms over 100 square feet	1 hour or provide automatic fire-extinguishing system
Storage rooms over 100 square feet	1 hour or provide automatic fire-extinguishing system
Group I-3 cells equipped with padded surfaces	1 hour
Group I-2 waste and linen collection rooms	1 hour
Waste and linen collection rooms over 100 square feet	1 hour or provide automatic fire-extinguishing system
Stationary lead-acid battery systems having a liquid capacity of more than 100 gallons used for facility standby power, emergency power or uninterrupted power supplies	1-hour fire barriers and floor/ceiling assemblies in Group B, F, H, M, S and U occupancies. 2-hour fire barriers and floor/ceiling assemblies in Group A, E, I and R occupancies

For SI: 1 square foot = 0.0929 m^2, 1 pound per square inch = 6.9 kPa, 1 British thermal unit = 0.293 watts, 1 horsepower = 746 watts, 1 gallon = 3.785 L.

a. Where an automatic fire-extinguishing system is provided, it need only be provided in the incidental use room or area.

Q: Is a laundry room located in a building of occupancies A, B, E, I-1, R-1 or R-2, which is available and usable by all building occupants, regarded as a space that is to be considered a storage room and therefore required to have the specified separation/protection in accordance with Table 508.2?

A: Yes. Laundry rooms over 100 square feet in area are required to have the specified separation/protection.

508.2.1 Occupancy classification. An incidental use area shall be classified in accordance with the occupancy of that portion of the building in which it is located or the building shall be classified as a mixed occupancy and shall comply with Section 508.3.

Q: Does Section 508.2.1 apply to storage or workshop rooms or other areas where such locations are considered a separate occupancy, either Use Group S or F, and where the mixed use groups are designed in accordance with Section 508.3?

A: No. Section 502.1 is intended to apply to specific occupancy areas that are considered a part of the main occupancy classification of the building. Under this circumstance, the building would not be subject to the code requirements related to occupancy Group S or F. Such areas generally represent a higher relative fire hazard than what is considered in establishing the main occupancy classification (A, B, E, I, R-1 and R-2) of the building, which is utilized to determine the applicable code requirements. Thus, the presence of the higher relative hazard is accounted for by requiring that area to be separated and/or protected as specified in the code.

However, such areas can be classified separately as to their occupancy use, and the building considered to be of mixed occupancy use. The presence of the higher relative fire hazard must be accounted for by protection features. The protection features result from considering such areas as one of the building's occupancy classifications. The application of Section 508.3 will result in certain building features being governed by code requirements commensurate with higher relative fire hazard of a storage or workshop area.

508.3 Mixed occupancies. Each portion of a building shall be individually classified in accordance with Section 302.1.

Where a building contains more than one occupancy group, the building or portion thereof shall comply with Sections 508.3.1, 508.3.2, 508.3.3 or a combination of these sections.

Exceptions:

1. Occupancies separated in accordance with Section 509.
2. Where required by Table 415.3.2, areas of Group H-1, H-2 or H-3 occupancies shall be located in a separate and detached building or structure.

Q: When a building that is classified other than Group S contains storage space, can the building be classified other than Group S and also be classified as mixed use and designed to comply with the provisions in Section 508.3 in lieu of the separation requirements for incidental use storage required by Table 508.2?

A: Yes. The code does not mandate compliance with Table 508.2 Incidental Use Areas if the building is designed as a mixed use. Table 508.2 is applicable to incidental use storage areas that are not considered mixed uses, and the storage areas are separated from the main occupancy as required in Table 508.2; however, such storage areas could be classified separately with regards to occupancy use and the building designed to comply with the provisions of Section 508.3 Mixed occupancies. The building could therefore be designed with either nonseparated (508.3.2) or separated (508.3.3) uses.

508.3.1 Accessory occupancies. Accessory occupancies are those occupancies subsidiary to the main occupancy of the building or portion thereof. Aggregate accessory occupancies shall not occupy more than 10 percent of the area of the story in which they are located and shall not exceed the tabular values in Table 503, without height and area increases in accordance with Sections 504 and 506 for such accessory occupancies.

Exceptions:

1. Accessory assembly areas having a floor area less than 750 square feet (69.7 m^2) are not considered separate occupancies.
2. Assembly areas that are accessory to Group E occupancies are not considered separate occupancies except when applying the assembly occupancy requirements of Chapter 11.
3. Accessory religious educational rooms and religious auditoriums with occupant loads of less than 100 are not considered separate occupancies.

Q: Can a building with a single occupancy classification contain other accessory areas or specific occupancy areas?

A: Yes. A building is permitted to contain more than one occupancy and still be given an occupancy classification that is consistent with the main occupancy. All specific occupancy areas as listed in Table 508.2 must be separated and/or protected as required by the code. Other nonhigh-hazard accessory areas must be supplemental to the main occupancy and cannot in aggregate area exceed either of the following: 10 percent of the area of the story in which it is located or 10 percent of the allowable area permitted by Sections 504 and 506 based on the accessory occupancy classification. When the specific occupancy areas are separated and protected as required and nonhigh-hazard accessory areas are limited in size as required, the building can be classified according to the main use group only.

Q: Can a Group I-2 occupancy care facility be an accessory occupancy area in a building of Type IIIB or VB construction in accordance with Section 508.3.1?

A: No. The code prohibits Group I-2 occupancies in buildings of unprotected combustible construction. This action was consistent with the 1984 report by the CABO Board for the Coordination of the Model Codes entitled, "Health Care Facilities – Use Group I-2 Institutional." The reason for the prohibition is the potential fire hazard of combustible construction and combustible concealed spaces, coupled with the lack of fire-resistance-rated elements when considered in the context of the "defend-in-place" philosophy utilized by the code for such occupancies.

Q: Is an accessory assembly use area no greater than 750 square feet permitted to exceed 10 percent of the area of the story on which it is located?

A: Yes. The requirements in Section 508.3.1, Exception 1 do not consider an accessory assembly use area that is equal to or less than 750 square feet to be a separate occupancy from the primary occupancy, and occupancy separation by fire barrier per Table 508.3.3 is not required. In addition, an accessory assembly use area that does not exceed 750 square feet is not limited to 10 percent of the area of the story on which it is located, as provisioned in Section 508.3.1. An example of this is a 750 square feet accessory assembly use area located on a floor level of 5,000 square feet. The 750 square feet accessory assembly use area occupies 15 percent of the 5,000 floor area, but because it does not exceed 750 square feet it is not considered a separate occupancy and fire-resistive separation is not requirement.

SECTION 509
SPECIAL PROVISIONS

509.5 Group R-2 buildings of Type IIIA construction. The height limitation for buildings of Type IIIA construction in Group R-2 shall be increased to six stories and 75 feet (22 860 mm) where the first floor construction above the basement has a fire-resistance rating of not less than 3 hours and the floor area is subdivided by 2-hour fire-resistance-rated fire walls into areas of not more than 3,000 square feet (279 m^2).

Q: Does the 3,000-square-foot fire area as referred to in Section 509.5 apply to all floors of a building or only floors located above the basement?

A: The 3,000-square-foot maximum fire area applies to all floors where Type IIIA construction is used. The purpose of this section is to permit buildings of Type IIIA construction to be built to a greater height than otherwise permitted. To compensate for the increased fuel load and fire potential of the building, it is the intent of the code to limit the size of the fire areas with fire walls to 3,000 square feet per story.

509.7 Open parking garage beneath Groups A, I, B, M and R. Open parking garages constructed under Groups A, I, B, M and R shall not exceed the height and area limitations permitted under Section 406.3. The height and area of the portion of the building above the open parking garage shall not exceed the limitations in Section 503 for the upper occupancy. The height, in both feet and stories, of the portion of the building above the open parking garage shall be measured from grade plane and shall include both the open parking garage and the portion of the building above the parking garage.

Q: Given a mixed use group building with open parking below another occupancy as described in Section 509.7, is suppression of the open parking area required when the occupancy above must have an automatic fire suppression system?

A: No. Section 509.7 provides for various exceptions to requirements that would otherwise apply under the general mixed use group approach. This unique approach recognizes the unusually low fire hazard potential of an open parking structure. The requirement for fire-resistance-rated separation between an open parking structure and the occupancy above, as well as other conditions set forth in Section 509.7, adequately address the safety concerns of placing the parking below the other occupancy. An automatic fire suppression system for the open parking area would not appreciably increase the level of safety and, therefore, need not be required in this situation.

CHAPTER 6

TYPES OF CONSTRUCTION

SECTION 602
CONSTRUCTION CLASSIFICATION

602.1 General. Buildings and structures erected or to be erected, altered or extended in height or area shall be classified in one of the five construction types defined in Sections 602.2 through 602.5. The building elements shall have a fire resistance rating not less than that specified in Table 601 and exterior walls shall have a fire-resistance rating not less than that specified in Table 602.

602.1.1 Minimum requirements. A building or portion thereof shall not be required to conform to the details of a type of construction higher than that type, which meets the minimum requirements based on occupancy even though certain features of such a building actually conform to a higher type of construction.

Q: Can a building contain more than one construction type?

A: Yes. Section 602.1 indicates that a building or structure shall be classified in one of the five permitted construction types. However, if the building contains multiple types of construction, Section 602.1.1 indicates that the building's designated type of construction shall be that of the lesser type unless all requirements for the higher type are fulfilled. For example, if a building is constructed of both Type IB and Type VB construction, the building must be classified as Type VB construction for purposes of the application of code requirements.

An acceptable alternative is to construct a fire wall dividing the structure, thus creating two separate buildings. Each building can then be classified separately as to its construction type.

Q: In buildings of Type I or Type II construction, does the code allow exterior nonload-bearing walls to be constructed of steel stud and gypsum board with an exterior fire-retardant-treated wood sheathing?

A: No. In buildings of Type I or Type II construction, the exterior wall system must have the required fire-resistance ratings, structural strength and weather resistance without the use of combustible materials. In this particular situation, both steel studs and gypsum board are adequate, but a noncombustible exterior weather covering must be provided. Combustible architectural trim is permitted per Section 1406.2.2 and foam plastic insulation per Section 2603.

Fire-retardant-treated wood is a combustible material which is allowed as indicated in Section 603.1.

Q: Can fire-retardant-treated wood be used where the code requires the use of noncombustible materials?

A: No. Fire-retardant-treated wood is a combustible material. The allowable use of fire-retardant-treated wood in noncombustible construction is limited as indicated in Section 603.1.

Q: Are exterior walls placed perpendicular to an interior lot line required to be fire-resistance-rated because of proximity to the lot line?

A: No. Exterior walls are required to be fire-resistance rated in accordance with Table 602, based on the fire separation distance as defined in Section 702.1, and the building's construction type and use group classification. The fire separation distance for the exterior walls in question is measured from the face of the wall. Therefore, the fire separation distance and the fire-resistance rating are not determined by a lot line located perpendicular to the wall.

Q: Is the elevator rail beam located within a 2-hour fire-resistance-rated elevator shaft enclosure required to have a fire-resistance rating?

A: No.

Q: Are steel beams that support only precast concrete stringers and/or a concrete stair landing required to have a fire-resistance rating if they fall completely within the 2-hour fire-resistance-rated stairway enclosure?

A: No. Structural members that support stairway landings, and stringers or elevator guide rails which are located within a fire-resistance-rated shaft enclosure and do not constitute part of the required shaft enclosure or provide structural support of the shaft enclosure, would not be required to have an individual fire-resistance rating.

Q: Is it the intent of Section 703.2 and Table 601 that only roof/ceiling assemblies tested in accordance with the American Society for Testing and Material (ASTM) Standard E 119 comply with the requirement for fire-resistance-rated roof construction in Table 601?

A: Yes.

Q: If so, how does a wood truss roof system comply with this provision?

A: It is the intent of the code that fire-resistance ratings of structural elements be determined in accordance with ASTM E 119 or an approved analytical method. In the case of a wood truss roof system, if there is no tested assembly for a 1-hour roof/ceiling assembly, some method of equivalency must be obtained. The code official can consider approval of an alternative method based on the use of a ceiling membrane from a fire-resistance-rated roof/ceiling or floor/ceiling assembly in which the ceiling membrane provides equivalent protection to the bottom chord of the truss. A ceiling membrane with the appropriate connections and support providing a finish rating of 60 minutes or more can be utilized to determine equivalency.

Q: Where a residential building is developed in a manner that establishes lot lines between adjacent dwelling units, can the separation between such units consist of two separate walls in lieu of a party wall?

A: Yes. Where lot lines occur between adjacent dwelling units, each unit, in effect, is a building on an individual lot and has a zero fire separation distance from the common lot line between adjacent units. The wall construction between units can therefore be composed of two independent wall assemblies, each complying with the requirements for an exterior wall with a zero fire separation distance. In Type V construction, the walls are permitted to be of combustible construction, and each wall must provide a 1-hour fire-resistance rating with no openings in accordance with Section 704. Alternatively, a single, noncombustible wall assembly could be used on the lot line. Such a wall, by definition, would be a party wall and must comply with the requirements of Section 707.

Q: Is an elevator hoistway required to be enclosed with a fire-resistance-rated assembly as specified for elevator shafts where the elevator is located within an atrium space?

A: No. Shaft enclosures for elevator hoistways are required to protect penetrations of floor assemblies created by the hoistway. Where an elevator hoistway is located within an atrium space, no penetrations of floor assemblies occur. The atrium is one continuous volume of space; therefore, a fire-resistance-rated enclosure of the hoistway within such space is not required.

Q: Is it acceptable to enclose a corridor by installing a 1-hour fire-resistance-rated assembly below a nonrated roof assembly within the corridor only, as an alternative to either protecting the entire roof/ceiling assembly or extending the fire-resistance-rated walls to the roof deck?

A: Yes. Where Section 1016.1 requires 1-hour fire-resistance-rated exit access corridors, the intent of the code is to provide a 1-hour fire-resistance-rated enclosure. An independent 1-hour fire-resistance-rated assembly below the roof assembly within the corridor would provide the enclosure protection required for the corridor. In this situation, the remainder of the building roof/ceiling assembly is not required to be fire-resistance rated.

602.3 Type III. Type III construction is that type of construction in which the exterior walls are of noncombustible materials and the interior building elements are of any material permitted by this code. Fire-retardant-treated wood framing complying with Section 2303.2 shall be permitted within exterior wall assemblies of a 2-hour rating or less.

Q: If fire-retardant-treated wood studs are used in the exterior walls of Type V construction, would this change the construction to Type III?

A: No. Section 602.3 requires exterior walls of Type III construction be of noncombustible materials. The *International Building Code* recognizes the increased resistance to fire afforded by fire-retardant-treated wood, but does not consider it to be noncombustible.

SECTION 603 COMBUSTIBLE MATERIAL IN TYPE I AND II CONSTRUCTION

603.1 Allowable materials. Combustible materials shall be permitted in buildings of Type I or Type II construction in the following applications and in accordance with Sections 603.1.1 through 603.1.3:

1. Fire-retardant-treated wood shall be permitted in:
   - 1.1. Nonbearing partitions where the required fire-resistance rating is 2 hours or less.
   - 1.2. Nonbearing exterior walls where no fire rating is required.
   - 1.3. Roof construction, including girders, trusses, framing and decking.

     Exception: In buildings of Type I construction exceeding two stories in height, fire-retardant-treated wood is not permitted in roof construction when the vertical distance from the upper floor to the roof is less than 20 feet (6096 mm).
2. Thermal and acoustical insulation, other than foam plastics, having a flame spread index of not more than 25.

   Exceptions:

   1. Insulation placed between two layers of noncombustible materials without an intervening airspace shall be allowed to have a flame spread index of not more than 100.

   2. Insulation installed between a finished floor and solid decking without intervening airspace shall be allowed to have a flame spread index of not more than 200.
3. Foam plastics in accordance with Chapter 26.
4. Roof coverings that have an A, B or C classification.
5. Interior floor finish and interior finish, trim and millwork such as doors, door frames, window sashes and frames.
6. Where not installed over 15 feet (4572 mm) above grade, show windows, nailing or furring strips and wooden bulkheads below show windows, including their frames, aprons and show cases.
7. Finished flooring applied directly to the floor slab or to wood sleepers that are fireblocked in accordance with Section 717.2.7.
8. Partitions dividing portions of stores, offices or similar places occupied by one tenant only and that do not establish a corridor serving an occupant load of 30 or more shall be permitted to be constructed of fire-retardant-treated wood, 1-hour fire-resistance-rated construction or of wood panels or similar light construction up to 6 feet (1829 mm) in height.
9. Stages and platforms constructed in accordance with Sections 410.3 and 410.4, respectively.
10. Combustible exterior wall coverings, balconies and similar projections and bay or oriel windows in accordance with Chapter 14.
11. Blocking such as for handrails, millwork, cabinets and window and door frames.
12. Light-transmitting plastics as permitted by Chapter 26.
13. Mastics and caulking materials applied to provide flexible seals between components of exterior wall construction.
14. Exterior plastic veneer installed in accordance with Section 2605.2.
15. Nailing or furring strips as permitted by Section 803.4.
16. Heavy timber as permitted by Note d to Table 601 and Sections 602.4.7 and 1406.3.
17. Aggregates, component materials and admixtures as permitted by Section 703.2.2.
18. Sprayed fire-resistant materials and intumescent and mastic fire-resistant coatings, determined on the basis of fire-resistance tests in accordance with Section 703.2 and installed in accordance with Section 1704.10 and 1704.11, respectively.
19. Materials used to protect penetrations in fire-resistance-rated assemblies in accordance with Section 712.
20. Materials used to protect joints in fire-resistance-rated assemblies in accordance with Section 713.
21. Materials allowed in the concealed spaces of buildings of Type I and II construction in accordance with Section 717.5.
22. Materials exposed within plenums complying with Section 602 of the *International Mechanical Code*.

Q: Are combustible plumbing fixtures, such as water closets and shower and bathtub enclosures, permitted to be installed in buildings of Type I or Type II construction?

A: Yes. It is not the intent of the *International Building Code* that the combustibility or noncombustibility of plumbing fixtures be a factor that determines the type of construction classification. The *International Plumbing Code* specifies the requirements for plumbing fixtures. There is no distinction made for plumbing fixtures among the various construction types. The combustible plumbing fixtures that are most prevalent must conform to one of the American National Standard Institute (ANSI) Z124 standards requiring fixtures to be impregnated with chemicals to retard the contributions that the fixtures would make to combustion.

Q: Are doors and windows that are constructed of combustible materials permitted in noncombustible types of construction?

A: Yes. Doors and windows are not among the elements required to be composed of noncombustible materials in Types I, II, III and IV construction, as described in Sections 602.2, 602.3 and 602.4. When a door or window is not required to be a protected opening, the code does not contain any requirements regulating the use of combustible versus noncombustible materials (see Table 704.8).

Q: Are untreated wood blocking or nailers used to support fixtures, railings, cabinets, interior and exterior finishes, etc., permitted within walls and partitions required to be of noncombustible construction?

A: Yes. Item 15 of Section 603.1 permits combustible nailers and blocking as stipulated in Section 803.4. Section 803.4.1 indicates that "furring strips not exceeding 1.75 inches" are permitted to be used in concrete or masonry construction for securing trim and finish. Although locating these combustible elements within noncombustible frame partitions is not specifically identified in this section, the presence of combustible nailers within noncombustible construction types, other than concrete and masonry, represents an equivalent circumstance. Therefore, it is the intent of the code to permit the use of combustible nailers and blocking within Types I and II construction.

Q: Do Items 4 and 15 of Section 603.1 allow the use of combustible wood structural panels or nailing/furring strips in the roof construction of noncombustible buildings for the purposes of fastening steep-slope or low-slope roof coverings?

A: Yes. In accordance with Section 602.2, the roofs of noncombustible buildings are required to be constructed of noncombustible materials. "Roof covering" is defined as the membrane covering the roof which provides weather resistance, fire resistance and appearance. Section 603.1 permits 22 building components to be of combustible materials within the roofs, floors and walls of buildings of noncombustible construction without altering the construction classification. As long as a noncombustible roof deck is provided as the structural element, foam plastic insulation, wood structural panels, nailing/furring strips and roof coverings may be applied.

Q: Can untreated wood studs be substituted for metal studs in a nonload-bearing interior partition of a Type I or II building?

A: No. Section 602.2 states that in buildings of Type I or II construction, the walls, partitions, structure elements, floors, ceilings, roofs and exits are to be composed of noncombustible materials. Exceptions are indicated by Section 603.1.

Q: Can a wood frame load-bearing wall be constructed at the Inside surface of a complying noncombustible exterior wall in Type III construction?

A: Yes. Section 602.3 permits interior load-bearing walls in Type III construction to be composed of combustible construction. Because the noncombustible exterior wall must develop its weather resistance, structural integrity and fire-resistance rating without relying on the wood frame wall, the code does not regulate the location of the wood frame load-bearing wall within the structure.

CHAPTER 7

FIRE-RESISTANCE-RATED CONSTRUCTION

SECTION 703
FIRE-RESISTANCE RATING AND FIRE TESTS

703.2 Fire-resistance ratings. The fire-resistance rating of building elements shall be determined in accordance with the test procedures set forth in ASTM E 119 or in accordance with Section 703.3. Where materials, systems or devices that have not been tested as part of a fire-resistance-rated assembly are incorporated into the assembly, sufficient data shall be made available to the building official to show that the required fire-resistance rating is not reduced. Materials and methods of construction used to protect joints and penetrations in fire-resistance-rated building elements shall not reduce the required fire-resistance rating.

> **Exception:** In determining the fire-resistance rating of exterior bearing walls, compliance with the ASTM E 119 criteria for unexposed surface temperature rise and ignition of cotton waste due to passage of flame or gases is required only for a period of time corresponding to the required fire-resistance rating of an exterior nonbearing wall with the same fire separation distance, and in a building of the same group. When the fire-resistance rating determined in accordance with this exception exceeds the fire-resistance rating determined in accordance with ASTM E 119, the fire exposure time period, water pressure, and application duration criteria for the hose stream test of ASTM E 119 shall be based upon the fire-resistance rating determined in accordance with this exception.

Q: When an interior wall is required to have a fire-resistance rating, such as an exit access corridor wall, must the wall be rated for an exposure from both sides?

A: Yes. The intent of the code is to require the fire-resistance rating of interior walls to be based on a fire exposure from either side. In the case of an exit access corridor, it is appropriate to expect the required fire performance from the corridor side, as well as the room side. Should the corridor become involved in a fire incident, the corridor wall would serve to restrict further spread of a fire into adjacent spaces.

703.2.3 Restrained classification. Fire-resistance-rated assemblies tested under ASTM E 119 shall not be considered to be restrained unless evidence satisfactory to the building official is furnished by the registered design professional showing that the construction qualifies for a restrained classification in accordance with ASTM E 119. Restrained construction shall be identified on the plans.

Q: When evaluating the fire-resistance rating of a floor or roof assembly, must the support condition of the test assembly (restrained versus unrestrained) be taken into consideration?

A: Yes. Section 703.2 references ASTM E 119 as the appropriate test standard for evaluating the fire-resistance rating of building components and assemblies. Included in the test parameters of the standard are two conditions of support: restrained and unrestrained. Many different floor and roof assemblies have been tested in accordance with ASTM E 119 under both restrained and unrestrained conditions. The intent of Section 703.2.3 is to require that the in-place construction of fire-resistance-rated assemblies be representative of the conditions under which the assembly was tested.

Q: Does the code require that a fire-resistance-rated assembly that has been tested in accordance with ASTM E 119 in the restrained condition of support be constructed to resist thermal expansion during fire exposure?

A: Yes. The intent of the code is that the in-place construction of fire-resistance-rated assemblies be representative of the conditions under which the assembly was tested. Conditions of restraint are part of the testing parameters of ASTM E 119 and therefore must be representative of the constructed assembly and supporting elements in the field.

Floor and roof assemblies and individual beams are considered restrained when the surrounding or supporting construction is capable of resisting thermal expansion throughout the range of anticipated temperatures encountered in a fire scenario. Engineering analysis and judgement are necessary to determine whether a fire-resistance-rated beam or assembly is adequately supported to resist the induced thermal expansion.

SECTION 704
EXTERIOR WALLS

704.8 Allowable area of openings. The maximum area of unprotected or protected openings permitted in an exterior wall in any story shall not exceed the values set forth in Table 704.8. Where both unprotected and protected openings are located in the exterior wall in any story, the total area of the openings shall comply with the following formula:

$$\frac{A}{a} + \frac{A_u}{a_u} \leq 1.0 \qquad \textbf{(Equation 7-2)}$$

where:

A = Actual area of protected openings, or the equivalent area of protected openings, A_e (see Section 704.7).

a = Allowable area of protected openings.

A_u = Actual area of unprotected openings.

a_u = Allowable area of unprotected openings.

Q: Do Section 704.8 provisions regulating the maximum areas of unprotected or protected openings in exterior walls apply regardless of whether or not an exterior wall is required to have a fire-resistance rating in accordance with Section 704.5 provisions?

A: Yes. Section 704.8 provisions are intended to apply to all exterior walls regardless of whether or not the walls are required to be fire-resistance rated by Section 704.5. These requirements are based on the concept that even nonfire-resistance-rated exterior walls provide some protection against exposure from adjacent buildings or stories, and that limitations on openings are, therefore, appropriate for both fire-resistance-rated and nonfire-resistance-rated walls.

704.10 Vertical exposure. For buildings on the same lot, opening protectives having a fire-protection rating of not less than $^3/_4$ hour shall be provided in every opening that is less than 15 feet (4572 mm) vertically above the roof of an adjoining building or adjacent structure that is within a horizontal fire separation distance of 15 feet (4572 mm) of the wall in which the opening is located.

Exception: Opening protectives are not required where the roof construction has a fire-resistance rating of not less than 1 hour for a minimum distance of 10 feet (3048 mm) from the adjoining building, and the entire length and span of the supporting elements for the fire-resistance-rated roof assembly has a fire-resistance rating of not less than 1 hour.

Q: A covered mall building constructed in accordance with Section 402 is attached to a multistory anchor store. Is it the intent of Section 704.10 to require exterior opening protectives for the windows in the anchor store that are adjacent to and located above the roof of the covered mall building?

A: No. Although the anchor store is not considered part of the covered mall building, it is the intent of Section 402.7.3.1 to permit unprotected openings between the anchor store and the pedestrian area of the mall. It would be inconsistent to allow unprotected openings between the mall area and the anchor store and, at the same time, require exterior opening protectives in windows located in the anchor store above the roof of the covered mall building.

Q: Where Table 704.8 prohibits any openings in an exterior wall due to fire separation distance, is it the intent of Section 704.10 to permit openings in the exterior wall that is located above the roof of an adjacent building?

A: No. Section 704.8 regulates the maximum area of openings (protected and unprotected) in exterior walls. These limitations are applicable to each and every story of all exterior walls. The conditions addressed in Section 704.10 are also subject to the requirements of Section 704.8. Thus, the exterior wall of a building is required to comply with Section 704.8 for determining the maximum amount of openings permitted whether or not there is an adjoining building or adjacent structure with a lower roof height. Note f to Table 704.8 provides the cross reference to Section 705.6.1 for exterior opening requirements above fire walls.

Q: When evaluating the opening protective requirements of Section 704.10, must the fire separation distances of both buildings be considered?

A: Yes. The hazard addressed in Section 704.10 is due to the fire exposure from an adjacent lower roof which is in close proximity to the exterior wall of the higher building. The requirements of this section are based on a 15-foot fire separation distance as defined in Section 702, whereas the requirements of previous editions of model codes were based on a 30-foot horizontal distance between adjacent buildings. The current requirements presume that the imaginary line or lot line is drawn equidistant between two adjacent buildings located no more than 30 feet apart.

The provisions of Section 704.10 are intended to be applied where the adjacent exterior walls of both buildings are required by Table 602 to be fire-resistance rated. This typically occurs where both buildings have a fire separation distance of 15 feet or less. The hazard addressed in Section 704.10 does not exist where one building is located at a fire separation distance of 15 feet or less and the adjoining building or adjacent structure is located at a fire separation distance of greater than 15 feet.

SECTION 705
FIRE WALLS

705.2 Structural stability. Fire walls shall have sufficient structural stability under fire conditions to allow collapse of construction on either side without collapse of the wall for the duration of time indicated by the required fire-resistance rating.

Q: Can a fire-resistance-rated floor/ceiling assembly be classified as a horizontal fire wall for the purpose of subdividing a multistory structure into separate buildings?

A: No. Section 705.2 states that fire walls shall have sufficient structural stability under fire conditions to allow collapse of construction on either side without collapse of the wall. Obviously, collapse of the lower portion of the building would result in collapse of the entire structure. Thus, a "horizontal fire wall" cannot meet the structural performance requirement of a fire wall.

SECTION 706
FIRE BARRIERS

706.5 Continuity of fire barrier walls. Fire barriers shall extend from the top of the floor/ceiling assembly below to the underside of the floor or roof slab or deck above and shall be securely attached thereto. Such fire barriers shall be continuous through concealed spaces, such as the space above a suspended ceiling. The supporting construction for fire barrier walls shall be protected to afford the required fire-resistance rating of the fire barrier supported, except for 1-hour fire-resistance-rated incidental use area separations as required by Table 508.2 in buildings of Type IIB, IIIB and VB construction. Hollow vertical spaces within a fire barrier shall be fireblocked in accordance with Section 717.2 at every floor level.

Exceptions:

1. The maximum required fire-resistance rating for assemblies supporting fire barriers separating tank storage as provided for in Section 415.6.2.1 shall be 2 hours, but not less than required by Table 601 for the building construction type.
2. Shaft enclosures shall be permitted to terminate at a top enclosure complying with Section 707.12.

Q: Section 706.5 states that the building structure supporting a fire barrier wall shall be fire-resistance-rated equal to the fire barrier wall being supported. A building may be designed using unprotected construction, such as Type IIB. Depending on the height of the building, a stair shaft enclosure may be constructed with 1-hour or 2-hour fire-resistance-rated fire barrier walls. The fire barrier walls enclosing the shaft can be tied to and supported by the building construction.

Is it the intent to require the supporting construction throughout the entire building to be protected with fire-resistance-rated construction equal to the fire barrier walls?

A: No. Only the building elements within the load path of that portion of the building construction where the shaft enclosure is located are subject to the provisions of Section 706.5 with regard to supporting construction. It is not necessary to provide fire-resistance-rated construction equal to the fire barrier walls for all floors and supporting construction throughout the entire building. The purpose for allowing fire-resistance-rated construction to be isolated to the specific building elements providing structural support for fire barrier walls, is to permit vertical shaft enclosures to be located within buildings of unprotected types of construction, such as IIB, IIIB, and VB, without requiring a change of construction type for the entire building.

Q: In a building of Type IIB construction, a shaft wall is designed to be supported by the floor slab at each floor level. Is the steel structure supporting each floor slab required to have a fire-resistance rating equal to that required for the shaft?

A: Yes. Sections 706.5 and 714.1 require construction supporting fire barriers to afford the required fire-resistance rating of the supported assemblies. If a shaft is to be supported by each floor/ceiling assembly, the assemblies must develop a fire-resistance rating not less than that of the supported shaft. In a steel frame building, this protection must be provided for all members supporting the floor deck within the bay in which the shaft is located, including the columns defining the limits of the bay.

706.6 Exterior walls. Where exterior walls serve as a part of a required fire-resistance-rated enclosure or separation, such walls shall comply with the requirements of Section 704 for exterior walls, and the fire-resistance-rated enclosure or separation requirements shall not apply.

Exception: Exterior walls required to be fire-resistance rated in accordance with Section 1014.5.1 for exterior egress balconies, Section 1020.1.4 for exit enclosures and Section 1023.6 for exterior exit ramps and stairways.

Q: Where adjacent dwelling units have different roof heights, is the 1-hour fire partition required to extend to the underside of the higher roof (see Section 708.1, Item1)?

A: No. The requirements of Section 708.1, Item 1 only apply to the common interface separation of one dwelling unit from another dwelling unit. The wall above the lower roof is not common to the two dwelling units and would be considered an exterior wall which must comply with the code requirements applicable to exterior walls for the type of construction. (Note: If a fire wall/party wall is provided where an interior lot line exists between the two dwelling units, the wall above the lower roof would be regulated in accordance with Section 705.6.1).

SECTION 707
SHAFT ENCLOSURES

707.2 Shaft enclosure required. Openings through a floor/ceiling assembly shall be protected by a shaft enclosure complying with this Section.

Exceptions:

1. A shaft enclosure is not required for openings totally within an individual residential dwelling unit and connecting four stories or less.
2. A shaft enclosure is not required in a building equipped throughout with an automatic sprinkler system in accordance with Section 903.3.1.1 for an escalator opening or stairway that is not a portion of the means of egress protected according to Item 2.1 or 2.2:
   - 2.1. Where the area of the floor opening between stories does not exceed twice the horizontal projected area of the escalator or stairway and the opening is protected by a draft curtain and closely spaced sprinklers in accordance with NFPA 13. In other than Groups B and M, this application is limited to openings that do not connect more than four stories.
   - 2.2. Where the opening is protected by approved power-operated automatic shutters at every penetrated floor. The shutters shall be of noncombustible construction and have a fire-resistance rating of not less than 1.5 hours. The shutter shall be so constructed as to close immediately upon the actuation of a smoke detector installed in accordance with Section 907.11 and shall completely shut off the well opening. Escalators shall cease operation when the shutter begins to close. The shutter shall operate at a speed of not more than 30 feet per minute (152.4 mm/s) and shall be equipped with a sensitive leading edge to arrest its progress where in contact with any obstacle, and to continue its progress on release therefrom.
3. A shaft enclosure is not required for penetrations by pipe, tube, conduit, wire, cable and vents protected in accordance with Section 712.4.
4. A shaft enclosure is not required for penetrations by ducts protected in accordance with Section 712.4. Grease ducts shall be protected in accordance with the *International Mechanical Code.*
5. In other than Group H occupancies, a shaft enclosure is not required for floor openings complying with the provisions for atriums in Section 404.
6. A shaft enclosure is not required for approved masonry chimneys where annular space protection is provided at each floor level in accordance with Section 717.2.5.
7. In other than Groups I-2 and I-3, a shaft enclosure is not required for a floor opening or an air transfer opening that complies with the following:
   - 7.1. Does not connect more than two stories.
   - 7.2. Is not part of the required means of egress system, except as permitted in Section 1020.1.
   - 7.3. Is not concealed within the building construction.
   - 7.4. Is not open to a corridor in Group I and R occupancies.
   - 7.5. Is not open to a corridor on nonsprinklered floors in any occupancy.
   - 7.6. Is separated from floor openings and air transfer openings serving other floors by construction conforming to required shaft enclosures.
   - 7.7. Is limited to the same smoke compartment.
8. A shaft enclosure is not required for automobile ramps in open and enclosed parking garages constructed in accordance with Sections 406.3 and 406.4, respectively.
9. A shaft enclosure is not required for floor openings between a mezzanine and the floor below.
10. A shaft enclosure is not required for joints protected by a fire-resistant joint system in accordance with Section 713.
11. A shaft enclosure shall not be required for floor openings created by unenclosed stairs or ramps in accordance with Exception 8 or 9 in Section 1020.1.
12. Floor openings protected by floor fire doors in accordance with Section 711.8.
13. Where permitted by other sections of this code.

Q: All openings are required to be enclosed or otherwise protected to prevent the spread of fire from story to story. Are each of the exceptions to be considered separately?

A: Yes. Each exception is to be considered separately and independently.

707.14.1 Elevator lobby. An enclosed elevator lobby shall be provided at each floor where an elevator shaft enclosure connects more than three stories. The lobby shall separate the elevator shaft enclosure doors from each floor by fire partitions equal to the fire-resistance rating of the corridor and the required opening protection. Elevator lobbies shall have at least one means of egress complying with Chapter 10 and other provisions within this code.

Exceptions:

1. Enclosed elevator lobbies are not required at the street floor, provided the entire street floor is equipped with an automatic sprinkler system in accordance with Section 903.3.1.1.
2. Elevators not required to be located in a shaft in accordance with Section 707.2 are not required to have enclosed elevator lobbies.
3. Where additional doors are provided at the hoistway opening in accordance with Section 3002.6. Such doors shall be tested in accordance with UL 1784 without an artificial bottom seal.
4. In other than Group I-3, and buildings having occupied floors located more than 75 feet (22 860 mm) above the lowest level of fire department vehicle access, enclosed elevator lobbies are not required where the building is protected by an automatic sprinkler system installed in accordance with Section 903.3.1.1 or 903.3.1.2.
5. Smoke partitions shall be permitted in lieu of fire partitions to separate the elevator lobby at each floor where the building is equipped throughout with an automatic sprinkler system installed in accordance with Section 903.3.1.1 or 903.3.1.2.
6. Enclosed elevator lobbies are not required where the elevator hoistway is pressurized in accordance with Section 707.14.2.

Q: Given a building on grade, without a basement, see Attachment No. 1 Multi-Story Building. Is the second story of this building the first story above the lowest level of fire department vehicular access?

A: No. When determining stories above the lowest level of fire department vehicular access, a level, or floor level, is not a story. A "level" is the horizontal plane that is part of a story, not the entire story height. A "story" is the vertical space between the upper surface of one floor level and the upper surface of the floor level next above or below.

The level of fire department vehicle access as shown in attached drawing of a Multi-Story Building is at elevation 0.0 feet, the first floor level of the building is also at elevation 0.0 feet; therefore, the level of fire department vehicle access and the first floor level of the building are at the same elevation. The first story of the building begins at elevation 0.0 feet (first floor) and extends to elevation 10.0 feet (second floor). The first story of the building is the first story above the level of fire department vehicle access.

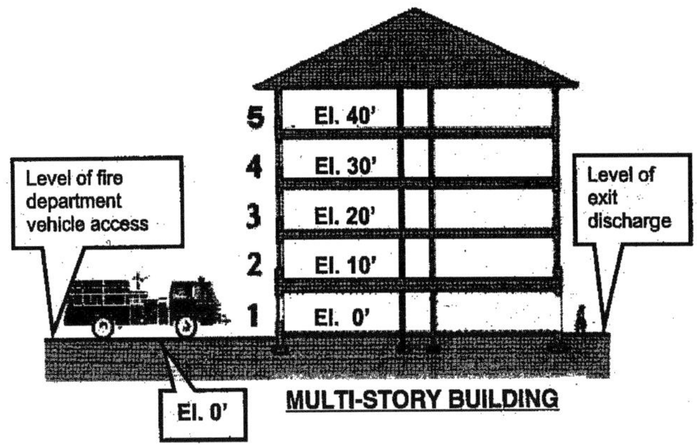

Attachment No. 1

SECTION 708
FIRE PARTITIONS

708.1 General. The following wall assemblies shall comply with this section:

1. Walls separating dwelling units in the same building.
2. Walls separating sleeping units in occupancies in Group R-1 hotel, R-2 and I-1 occupancies.
3. Walls separating tenant spaces in covered mall buildings as required by Section 402.7.2.
4. Corridor walls as required by Section 1017.1.
5. Elevator lobby separation as required by Section 707.14.1.
6. Residential aircraft hangars.

Q: A mixed occupancy building contains the following occupancies: Group R-1 Hotel, Group M, and Group B. The building is designed in compliance with Section 508.3.2 Nonseparated Uses. Is a fire partition required between a Group R-1 guest room and the adjacent Group M or Group B occupancy?

A: No. A fire partition is only required to separate the individual guest rooms within the Group R-1 occupancy. Neither fire partitions nor fire barriers are required to separate the guest rooms from the other occupancy classifications within the building because the building is designed in compliance with the requirements of Section

508.3.2 Nonseparated Uses where separation of occupancies (fire barrier, see Section 706.3.8) is not required.

708.3 Fire-resistance rating. Fire partitions shall have a fire-resistance rating of not less than 1 hour.

Exceptions:

1. Corridor walls as permitted by Table 1017.1.
2. Dwelling and sleeping unit separations in buildings of Type IIB, IIIB and VB construction shall have fire-resistance ratings of not less than $^1/_2$ hour in buildings equipped throughout with an automatic sprinkler system in accordance with Section 903.3.1.1.

Q: Are the requirements of Section 708.3 regarding separation between multiple single-family dwelling units affected where lot lines occur between dwelling units?

A: Yes. This provision is based on the premise that the dwelling units are considered collectively as one building on one lot. Where interior lot lines occur between individual dwelling units, Section 708.3 requirements for interior fire partitions are not applicable. A party wall constructed on the interior lot line must comply with Section 705 requirements. Alternatively, two separate and independent walls can be utilized; each of which must conform to Section 704 requirements for exterior walls with a zero fire separation distance.

Q: Side-by-side dwelling units in a building containing a residential occupancy with multiple single-family dwellings are offset such that only a portion of the wall that separates adjacent units is shared by both dwelling units. Are the opposing exterior walls of the adjacent units required to be fire-resistance rated?

A: Yes. Section 708.3 requires adjacent dwelling units to be separated with a 1-hour fire separation assembly in accordance with Section 708. Section 708.5 requires exterior walls that serve as part of the required separation to comply with the requirements in Section 704.

As the width of the open airspace between opposing exterior walls becomes smaller, the hazard of fire spreading from one dwelling unit to another increases, eventually resulting in the same hazard presented by side-by-side dwelling units. In such a case, opposing exterior walls must provide the same level of protection as the 1-hour fire separation assembly required to separate the interior portion of the wall held in common by the adjacent dwelling units. Conversely, as the width of the open airspace increases, the risk of fire spread between adjacent units is reduced, eventually reaching the point where no fire-resistance rating is warranted.

The exterior wall fire-resistance ratings and opening limitations provided by Table 602 recognize that the hazards associated with fire spread between adjacent dwelling units are a function of radiant heat exposure regardless of the presence of legal lot lines or imaginary lines. The building designer must establish imaginary lines between the adjacent dwelling units (similar to separate and independent buildings located on the same lot) to establish fire separation distances for the exterior walls in question. These fire separation distances must be used along with Tables 602 and 704.8 to determine the applicable fire-resistance ratings and opening limitations.

SECTION 711 HORIZONTAL ASSEMBLIES

711.3.3 Unusable space. In 1-hour fire-resistance-rated floor construction, the ceiling membrane is not required to be installed over unusable crawl spaces. In 1-hour fire-resistance-rated roof construction, the floor membrane is not required to be installed where unusable attic space occurs above.

Q: If a crawl space located below a fire-resistance-rated floor is readily accessible for storage, would Section 711.3.3 allow the ceiling membrane to be omitted in the crawl space?

A: No. The crawl space described above is accessible for storage, and thus not considered an unusable space. The use of the crawl space would introduce a hazard from below the floor. Therefore, Section 711.3.3 would not permit the ceiling membrane to be omitted. The ceiling membrane could be omitted only if the space is unusable and inaccessible for storage purposes.

711.4 Continuity. Assemblies shall be continuous without openings, penetrations or joints except as permitted by this section and Sections 707.2, 712.4, 713 and 1020.1. Skylights and other penetrations through a fire-resistance-rated roof deck or slab are permitted to be unprotected, provided that the structural integrity of the fire-resistance-rated roof construction is maintained. Unprotected skylights shall not be permitted in roof construction required to be fire-resistance rated in accordance with Section 704.10. The supporting construction shall be protected to afford the required fire-resistance rating of the horizontal assembly supported.

Q: Does Section 704.10 require opening protectives for exterior wall openings where the 1-hour fire-resistance-rated roof assembly of an adjacent lower building contains nonfire-resistance-rated skylights?

A: Yes. Section 711.4 specifically allows skylights in fire-resistance-rated assemblies, except where the hazards indicated in Section 704.10 are present. Where openings in exterior walls are subjected to vertical fire exposure, the unique protection requirements of Section 704.10 are intended to minimize the hazard. In this situation alone, the provision of a fire-resistance-rated barrier to restrict fire spread between adjacent but separate buildings becomes a necessity. A 1-hour fire-resistance-rated roof provides sufficient protection so long as the barrier is not compromised by unprotected penetrations that would expose the adjacent exterior wall openings to a fire condition. Accordingly, where the fire-resistance rating of the roof construction required by Section 704.10 is not maintained, as is the case where nonfire-resistance-rated skylights are used, the adjacent exterior wall openings must be protected.

SECTION 714
FIRE-RESISTANCE RATING OF STRUCTURAL MEMBERS

714.3 Embedments and enclosures. Pipes, wires, conduits, ducts or other service facilities shall not be embedded in the required fire protective covering of a structural member that is required to be individually encased.

Q: Does Section 714.3 prohibit the complete enclosure of pipes, wires, conduits, ducts, etc., within the empty spaces formed between structural members and the assembly's required fire-resistance-rated protective covering?

A: No. Piping, conduit and other service facilities can be installed in the hollow spaces of a fire-resistance-rated structural member assembly, provided that they are not embedded directly in the protective covering. Any penetrations of the protective covering assembly must not violate the integrity of the fire-resistance rating. (Note: Items that might significantly increase the ambient temperature within the assembly, such as steam piping, should not be permitted.)

SECTION 715
OPENING PROTECTIVES

715.4 Fire door and shutter assemblies. Approved fire door and fire shutter assemblies shall be constructed of any material or assembly of component materials that conforms to the test requirements of Section 715.4.1, 715.4.2 or 715.4.3 and the fire-protection rating indicated in Table 715.4. Fire door assemblies and shutters shall be installed in accordance with the provisions of this section and NFPA 80.

Exceptions:

1. Labeled protective assemblies that conform to the requirements of this section or UL 10A, UL 14B and UL 14C for tin-clad fire door assemblies.
2. Floor fire door assemblies in accordance with Section 711.8.

Q: Are mail slots and other similar openings allowed in doors that are required by Section 715.4 to have a specific fire protection rating?

A: Yes. Section 715.4.1 states that fire door assemblies shall be constructed of any material or assembly of component materials complying with NFPA 252 test requirements. Therefore, the code does not prohibit the inclusion of mail slots or other such openings in a rated fire door assembly but does require that such devices be included in the tested fire door assembly.

715.5.3 Wired glass. Steel window frame assemblies of 0.125-inch (3.2 mm) minimum solid section or of not less than nominal 0.048-inch-thick (1.2 mm) formed sheet steel members fabricated by pressing, mitering, riveting, interlocking or welding and having provision for glazing with $^1/_4$-inch (6.4 mm) wired glass where securely installed in the building construction and glazed with $^1/_4$-inch (6.4 mm) labeled wired glass shall be deemed to meet the requirements for a $^3/_4$-hour fire window assembly. Wired glass panels shall conform to the size limitations set forth in Table 715.5.3.

Q: Can the fire partition enclosures of corridors required to be fire-resistance-rated by Table 1017.1 contain an unlimited number of $^1/_4$-inch wired glass panels?

A: Yes. Section 715.5.3 permits 1-hour fire-resistance-rated fire separation assemblies to be constructed of an unlimited number of wired glass panels so long as the maximum size of each panel does not exceed the limitations for a $^3/_4$-hour door. Because fire barriers provide a higher degree of protection than fire partitions, it is consistent to apply the same limitations to a fire partition.

Accordingly, the entire wall may be constructed of wired glass panels, provided that each panel does not exceed the size limitations for an opening with a $^3/_4$-hour fire protection rating. The maximum area of each panel is limited to 1,296 square inches, and the maximum height and width is not permitted to exceed 54 inches.

CHAPTER 8

INTERIOR FINISHES

SECTION 803
WALL AND CEILING FINISHES

803.1 General. Interior wall and ceiling finishes shall be classified in accordance with ASTM E 84. Such interior finish materials shall be grouped in the following classes in accordance with their flame spread and smoke-developed indexes.

Class A: Flame spread 0-25; smoke-developed 0-450.

Class B: Flame spread 26-75; smoke-developed 0-450.

Class C: Flame spread 76-200; smoke-developed 0-450.

Exception: Materials, other than textiles, tested in accordance with Section 803.2.

Q: Do the interior finish flame spread limitations apply to doors?

A: No. Section 803 sets forth flame spread limitations of interior finish and wainscotting, paneling or other finish applied structurally or for acoustical treatment, decoration or similar purposes to walls and ceilings. Doors are not intended to be regulated under the interior finish requirements.

Q: Do the interior finish requirements apply to panel materials, such as pegboards, that are attached to walls that otherwise fully comply with the code?

A: Yes. A material permanently attached to walls and ceilings for structural, decoration or similar purposes would be regulated by the interior finish and trim requirements of Section 803.

803.5 Interior finish requirements based on group. Interior wall and ceiling finish shall have a flame spread index not greater than that specified in Table 803.5 for the group and location designated. Interior wall and ceiling finish materials, other than textiles, tested in accordance with NFPA 286 and meeting the acceptance criteria of Section 803.2.1, shall be permitted to be used where a Class A classification in accordance with ASTM E 84 is required.

Q: Exterior exit stairways complying with Section 1023 are used as an element of the means of egress. The exterior exit stairways connect to an open-ended corridor. Is the open-ended corridor that leads to the exterior exit stairways considered interior and required to comply with the interior finish requirements?

A: Yes. The open-ended corridor is an enclosed exit access component that defines and provides a path of egress travel to an exit. Even though the open-ended corridor is open on the ends, it is enclosed within the interior of the building by the ceiling, floor, and walls.

[F] SECTION 806
DECORATIVE MATERIALS AND TRIM

[F] 806.1.2 Combustible decorative materials. The permissible amount of decorative materials meeting the flame propagation performance criteria of NFPA 701 shall not exceed 10 percent of the aggregate area of walls and ceilings.

Exceptions:

1. In auditoriums in Group A, the permissible amount of decorative material meeting the flame propagation performance criteria of NFPA 701 shall not exceed 50 percent of the aggregate area of walls and ceiling where the building is equipped throughout with an automatic sprinkler system in accordance with Section 903.3.1.1 and where the material is installed in accordance with Section 803.4.
2. The amount of fabric partitions suspended from the ceiling and not supported by the floor in Group B and M occupancies shall not be limited.

Q: Is the area of flame-resistant decorative material limited to 10 percent of the total wall and ceiling areas of the room where materials are installed?

A: Yes. Section 806.1.2 limits the amount of flame-resistant decorative hangings to 10 percent of the total wall and ceiling area without indicating whether it is a room-by-room, tenant space or story-by-story consideration. The purpose of this section is to limit the concentrated amount of such materials that would be exposed to a single fire environment. As a result, it is reasonable to assume that the text is referring to a room or space similar to the manner in which Section 806.5 regulates interior trim. This would then eliminate the possibility of using areas located outside the probable fire environment and, in fact, limit the potential concentration of fire fuel in a single area. The use of the word "total" is intended to deal with areas considered "in the aggregate."

[F] 806.5 Interior trim. Material, other than foam plastic used as interior trim shall have a minimum Class C flame spread and smoke-developed index when tested in accordance with ASTM E 84, as described in Section 803.1. Combustible trim, excluding handrails and guardrails, shall not exceed 10 percent of the aggregate wall or ceiling area in which it is located.

Q: Do Sections 803.1 and 806.5 prohibit the use of interior trim that has a smoke-developed rating greater than 450 where the area of such trim does not exceed 10 percent of the aggregate wall and ceiling areas?

A: Yes. The smoke-developed limitation specified in Section 803.1 is applicable to all interior finish and trim materials except textile wall coverings, which comply with Section 803.6. Section 806.5 permits a flame spread rating of up to Class C if the area of such trim is less than 10 percent of the aggregate wall and ceiling area but does not modify or exempt the smoke-developed limitation.

Q: When Table 803.5 requires Class A or B interior finishes, does the table also permit the use of Class B or C interior trim that does not exceed 10 percent of the aggregate wall and ceiling area in required exits and passageways, as well as in corridors providing exit access?

A: Yes. The use of Class A, B or C materials for trim in accordance with Section 806.5 is applicable to all locations within a building.

TABLE 803.5
INTERIOR WALL AND CEILING FINISH REQUIREMENTS BY OCCUPANCY[k]

GROUP	SPRINKLERED[l]			NONSPRINKLERED		
	Exit enclosures and exit passageways[a,b]	Corridors	Rooms and enclosed spaces[c]	Exit enclosures and exit passageways[a,b]	Corridors	Rooms and enclosed spaces[c]
A-1 & A-2	B	B	C	A	A[d]	B[e]
A-3[f], A-4, A-5	B	B	C	A	A[d]	C
B, E, M, R-1, R-4	B	C	C	A	B	C
F	C	C	C	B	C	C
H	B	B	C[g]	A	A	B
I-1	B	C	C	A	B	B
I-2	B	B	B[h, i]	A	A	B
I-3	A	A[j]	C	A	A	B
I-4	B	B	B[h, i]	A	A	B
R-2	C	C	C	B	B	C
R-3	C	C	C	C	C	C
S	C	C	C	B	B	C
U	No restrictions			No restrictions		

For SI: 1 inch = 25.4 mm, 1 square foot = $0.0929 m^2$.

a. Class C interior finish materials shall be permitted for wainscotting or paneling of not more than 1,000 square feet of applied surface area in the grade lobby where applied directly to a noncombustible base or over furring strips applied to a noncombustible base and fireblocked as required by Section 803.4.1.

b. In exit enclosures of buildings less than three stories in height of other than Group I-3, Class B interior finish for nonsprinklered buildings and Class C interior finish for sprinklered buildings shall be permitted.

c. Requirements for rooms and enclosed spaces shall be based upon spaces enclosed by partitions. Where a fire-resistance rating is required for structural elements, the enclosing partitions shall extend from the floor to the ceiling. Partitions that do not comply with this shall be considered enclosing spaces and the rooms or spaces on both sides shall be considered one. In determining the applicable requirements for rooms and enclosed spaces, the specific occupancy thereof shall be the governing factor regardless of the group classification of the building or structure.

d. Lobby areas in Group A-1, A-2 and A-3 occupancies shall not be less than Class B materials.

e. Class C interior finish materials shall be permitted in places of assembly with an occupant load of 300 persons or less.

f. For places of religious worship, wood used for ornamental purposes, trusses, paneling or chancel furnishing shall be permitted.

g. Class B material is required where the building exceeds two stories.

h. Class C interior finish materials shall be permitted in administrative spaces.

i. Class C interior finish materials shall be permitted in rooms with a capacity of four persons or less.

j. Class B materials shall be permitted as wainscotting extending not more than 48 inches above the finished floor in corridors.

k. Finish materials as provided for in other sections of this code.

l. Applies when the exit enclosures, exit passageways, corridors or rooms and enclosed spaces are protected by a sprinkler system installed in accordance with Section 903.3.1.1 or 903.3.1.2.

CHAPTER 9

FIRE PROTECTION SYSTEMS

SECTION 903
AUTOMATIC SPRINKLER SYSTEMS

[F] 903.2.1 Group A. An automatic sprinkler system shall be provided throughout buildings and portions thereof used as Group A occupancies as provided in this section. For Group A-1, A-2, A-3 and A-4 occupancies, the automatic sprinkler system shall be provided throughout the floor area where the Group A-1, A-2, A-3 or A-4 occupancy is located, and in all floors between the Group A occupancy and the level of exit discharge. For Group A-5 occupancies, the automatic sprinkler system shall be provided in the spaces indicated in Section 903.2.1.5.

Q: Do the automatic sprinkler requirements in Section 903.2.1, based on the fire area of an occupancy, apply to buildings without surrounding exterior walls?

A: Yes. The entire area of a single floor of a building or structure must be considered when determining the automatic sprinkler requirements based on fire area. The entire area of a single floor of a building or structure may be considered a single fire area, or a portion of the floor area within a single story of a building or structure may be considered a fire area when separated in accordance with Section 706 of the *International Building Code*.

Where no surrounding exterior walls are provided along the perimeter of a building, the building area is used to identify and determine applicable fire areas. Section 502.1 of the *International Building Code* defines "Building area" as: "The area included within surrounding exterior walls (or exterior walls and fire walls) exclusive of vent shafts and courts. Areas of the building not provided with surrounding walls shall be included in the building area if such areas are included within the horizontal projection of the roof or floor above."

Q: Do the automatic sprinkler requirements in Section 903.2.1, based on the fire area of an occupancy, apply to buildings without surrounding exterior walls if the building perimeter is enclosed with a barrier, such as fencing, rails or other obstruction that restricts exiting from any location along the perimeter of the building?

A: Yes. Means of egress is not a determining factor in identifying a fire area.

[F] 903.2.7 Group R. An automatic sprinkler system installed in accordance with Section 903.3 shall be provided throughout all buildings with a Group R fire area.

Q: Do the provisions of Section 903.2.7 in the *International Fire Code* require automatic sprinkler systems to be installed in all one- and two-family dwellings?

A: No. The provisions of the *International Fire Code* are applicable only to one- and two-family dwelling buildings constructed under the *International Building Code* and classified as Group R-3 by that code. They do not apply to detached one- and two-family dwellings and multiple single-family dwellings (townhouses) not more than three stories in height with a separate means of egress constructed under the *International Residential Code*.

The ICC Board of Directors, in originally approving the development of the *International Residential Code*, indicated that it is to be a stand-alone code for the construction of detached one- and two-family dwellings and multiple single-family dwellings (townhouses) not more than three stories in height with a seperate means of egress. The provisions for new construction that affect these type buildings are regulated exclusively by the *International Residential Code* and are not regulated by any other *International Code*. The exception to Section 101.2 of the *International Building Code* stipulates that a detached one- and two-family dwellings and multiple single-family dwellings (townhouses) that are not more than three stories above grade plane in height shall comply with the *International Residential Code*. Likewise, Section R101.2, of the *International Residential Code* states that the provisions of the *International Residential Code* shall apply to the construction of detached one- and two-family dwellings and multiple single-family dwellings (townhouses) that are not more than three stories above grade in height with a separate means of egress and their accessory structures. Therefore, the construction of detached one- and two-family dwellings and townhouses within the scope of the *International Residential Code* is not required to comply with the provisions of either the *International Building Code* or the *International Fire Code*.

Where a dwelling or townhouse exceeds the allowable height in stories, does not provide individual egress for each dwelling unit, or does not conform to the prescriptive provisions of the *International Residential Code*, the structures are then beyond the scope of that code and its provisions cannot be applied. The building must then meet the provisions of the *International Building Code* for Group R-3 as well as the applicable provisions of the *International Fire Code*.

[F] 903.2.10.1 Stories and basements without openings. An automatic sprinkler system shall be installed throughout every story or basement of all buildings where the floor area exceeds 1,500 square feet (139.4 m^2) and where there is not provided at least one of the following types of exterior wall openings:

1. Openings below grade that lead directly to ground level by an exterior stairway complying with Section 1009 or an outside ramp complying with Section 1010. Openings shall be located in each 50 linear feet (15 240 mm), or fraction thereof, of exterior wall in the story on at least one side.
2. Openings entirely above the adjoining ground level totaling at least 20 square feet (1.86 m^2) in each 50 linear feet (15 240 mm), or fraction thereof, of exterior wall in the story on at least one side.

Q: An automotive service shop has below-grade service areas where employees perform oil changes and other minor maintenance services. The below-grade areas are open to the grade-level service bays, thus providing access to the underside of the vehicles without requiring the vehicle to be lifted into the air. Insofar as this below-grade space has no openings directly to the exterior, is it regulated as a windowless story and required to be equipped with an automatic fire suppression system in accordance with Section 903.2.10.1?

A: No. Due to the openness between the adjacent service levels, the below-grade area would be more appropriately regulated similar to a mezzanine rather than a story. A mezzanine is not regulated as a separate story but rather as part of the same story that it serves.

Therefore, if the below-grade service level is in compliance with the applicable provisions of Section 505, the windowless story provisions of Section 903.2.10.1 would be evaluated based on the exterior wall openings of the main level and not the service mezzanine below. The direct interconnections provided between two adjacent floor levels by multiple service openings provide access to the lower service area for fire-fighting and rescue operations. Due to the openness of the below-grade service level, it would not be regulated as a windowless story.

Q: A building is provided with openings in the exterior wall in lieu of the automatic fire suppression system in accordance with Section 903.2.10.1. Is the spacing between the jambs of adjacent openings in the exterior wall permitted to exceed 50 feet?

A: No. Section 903.2.10.1 requires that either exterior stairways, outside ramps or above-ground openings at least 20 square feet in size be located in each 50 lineal feet or fraction thereof of exterior walls. The required openings must be distributed such that the lineal distance between adjacent openings does not exceed 50 feet. The distribution of openings provides fire fighters with ready access to the interior of the building as well as multiple locations to vent smoke from the story in a fire situation.

If the openings in the exterior wall are located without regard to the location of adjacent openings, it is possible that segments of the exterior wall are not provided with the required access to the interior of the building for firefighting purposes. Any arrangement of required stairways, ramps or openings that results in a portion of the wall 50 feet or more in length with no openings to the exterior does not meet the intent of the code that access be provided in each 50 lineal feet.

[F] 903.3.1.2 NFPA 13R sprinkler systems. Where allowed in buildings of Group R, up to and including four stories in height, automatic sprinkler systems shall be installed throughout in accordance with NFPA 13R.

Q: Is an NFPA 13R sprinkler system permitted in a Group R occupancy where four stories of the building are located above grade plane and one story is located below grade plane?

A: Yes. The definition of "Building height" is the vertical distance from grade plane to the average height of the highest roof surface which establishes the grade plane as the reference point for determining building height and number of stories. Section 503 and Table 503 of the *International Building Code* establish the maximum building height and number of stories permitted for all occupancies above the grade plane. The maximum number of stories and building height in Table 503 of the *International Building Code* can be modified as permitted by Section 504.2 of the *International Building Code* when a sprinkler system is installed throughout in accordance with NFPA 13R. With the modification in Section 504.2, the building above the grade plane can be four stories provided the overall building height does not exceed 60 feet measured to the average height of the highest roof surface. The type of occupancy permitted in the basement, including residential, is not controlled by the *International Fire Code* provided the basement is separated from the residential portions of the building in accordance with the mixed-use provisions of the *International Building Code*. Group S-2 occupancies can also be installed in the lower level of a Group R occupancy if they comply with the requirements of Section 508 of the *International Building Code*. The lower level or basement of the building must comply with the *International Building Code* definition for basement.

[F] 903.4.2 Alarms. Approved audible devices shall be connected to every automatic sprinkler system. Such sprinkler water-flow alarm devices shall be activated by water flow equivalent to the flow of a single sprinkler of the smallest orifice size installed in the system. Alarm devices shall be provided on the exterior of the building in an approved location. Where a fire alarm system is installed, actuation of the automatic sprinkler system shall actuate the building fire alarm system.

Q: Is the additional, supplemental interior sprinkler system alarm, which is required in Section 903.4.2, intended for general building evacuation?

A: No. The system alarm located inside the building is intended to alert occupants that water is flowing in the sprinkler system. The code does not require this to be a general evacuation alarm.

SECTION 905 STANDPIPE SYSTEMS

[F] 905.3.1 Building height. Class III standpipe systems shall be installed throughout buildings where the floor level of the highest story is located more than 30 feet (9144 mm) above the lowest level of fire department vehicle access, or where the floor level of the lowest story is located more than 30 feet (9144 mm) below the highest level of fire department vehicle access.

Exceptions:

1. Class I standpipes are allowed in buildings equipped throughout with an automatic sprinkler system in accordance with Section 903.3.1.1 or 903.3.1.2.
2. Class I manual standpipes are allowed in open parking garages where the highest floor is located not more than 150 feet (45 720 mm) above the lowest level of fire department vehicle access.
3. Class I manual dry standpipes are allowed in open parking garages that are subject to freezing temperatures, provided that the hose connections are located as required for Class II standpipes in accordance with Section 905.5.
4. Class I standpipes are allowed in basements equipped throughout with an automatic sprinkler system.
5. In determining the lowest level of fire department vehicle access, it shall not be required to consider:
   - 5.1. Recessed loading docks for four vehicles or less; and
   - 5.2. Conditions where topography makes access from the fire department vehicle to the building impractical or impossible.

Q: Can a manual wet type water supply system be used in a Class I standpipe in a building that is not classified as a high-rise building?

A: Yes. Exception 1 to Section 905.3.1 permits the use of a Class I standpipe system in lieu of a Class III standpipe system when the building is equipped throughout with an automatic sprinkler system. However, Exception 1 does not stipulate a specific type of standpipe that must be used. The exception simply permits the use of a Class I standpipe. NFPA 14, as referenced in Section 905.2, provides specific design and installation requirements. A Class I standpipe system is intended for use by fire departments and trained personnel. A manual wet standpipe system requires the use of fire department equipment to pump water into the system to meet the demands put on the system.

SECTION 907 FIRE ALARM AND DETECTION SYSTEMS

[F] 907.2.4 Group F. A manual fire alarm system shall be installed in Group F occupancies that are two or more stories in height and have an occupant load of 500 or more above or below the lowest level of exit discharge.

Exception: Manual fire alarm boxes are not required when the building is equipped throughout with an automatic sprinkler system and the notification appliances will activate upon sprinkler water flow.

Q: Given a building on grade, without a basement, see Attachment No. 1 Multi-Story Building. Is the second story of this building the first story above the lowest level of exit discharge?

A: No. When determining stories above the lowest level of exit discharge, a level, or floor level, is not a story. A level is the horizontal plane that is part of a story, not the entire story height. A story is the vertical space between the upper surface of one floor level and the upper surface of the floor level next above or below.

The level of exit discharge as shown in the attached drawing of a Multi-Story Building is at elevation 0.0 feet, the first floor level of the building is also at elevation 0.0 feet; therefore, the level of exit discharge and the first floor level of the building are at the same elevation. The first story of the building begins at elevation 0.0 feet (first floor) and extends to elevation 10.0 feet (second floor).

The first story of the building is the first story above the level of exit discharge.

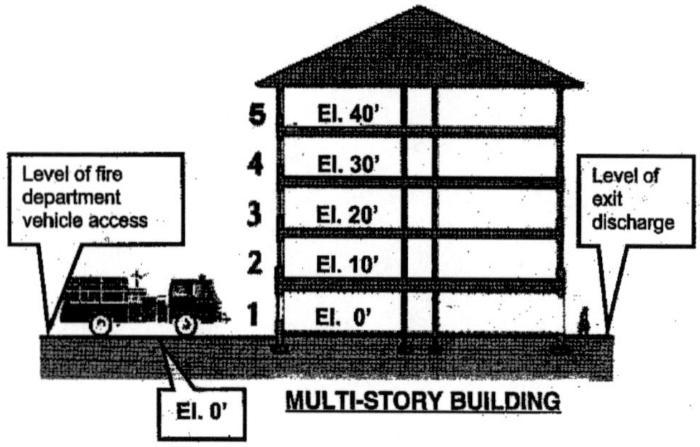

Attachment No. 1

Q: Is it the intent for the provisions of Section 907.2.4 to require a manual fire alarm system in a Group F occupancy where the occupant load of the first story of a multi-story building is 500 or more, and the first story is located at the lowest level of exit discharge?

A: No. The provisions of Section 907.2.4 do not require a manual fire alarm system in a Group F occupancy where the first story is the lowest level of exit discharge and the occupants have access to exit and exit discharge without being required to travel up or down an exit stairway. The provisions of Section 907.2.4 only apply to buildings two or more stories in height; they do not apply to a single-story building. A manual fire alarm system is required only when 500 or more occupants must travel up or down an exit stairway for no less than one story.

ICC Committee Interpretation 25-03, *International Fire Code* Section 907.2.4, issued May 11, 2004 addresses how to identify the location of stories relative to the level of exit discharge, and is not intended to address or clarify the specific scoping requirements for manual fire alarm systems in Group F occupancies.

This provision for a manual fire alarm system has its origin in the *Standard Building Code*, and used the term "above or below the street floor level," which was not defined. When the provision for a manual fire alarm system was brought forward into the working draft of the 2000 *International Building Code,* the term "above or below the level of exit discharge" was adopted; the original term "above or below the street floor level," used in the *Standard Building Code* was dropped. It was not the intent of the provisions in the *Standard Building Code* to require a manual fire alarm system when the minimum occupant load is located at or on the street floor level or on the "level of exit discharge."

In addition, "level of exit discharge" became a defined term in the *International Building Code* and no correlation was made to address the difference in the meaning and use of the two terms. As defined in the *International Building Code*, the "Level of exit discharge" is: "The horizontal plane located at the point at which an exit terminates and an exit discharge begins."

[F] 907.2.6 Group I. A manual fire alarm system shall be installed in Group I occupancies. An electrically supervised, automatic smoke detection system shall be provided in accordance with Sections 907.2.6.1 and 907.2.6.2.

> **Exception:** Manual fire alarm boxes in resident or patient sleeping areas of Group I-1 and I-2 occupancies shall not be required at exits if located at all nurses' control stations or other constantly attended staff locations, provided such stations are visible and continuously accessible and that travel distances required in Section 907.3.1 are not exceeded.

Q: Is an electrically supervised, smoke detection system required throughout a Group I-2 occupancy (hospital)?

A: No. Section 907.2.6 limits the need for electrically supervised, automatic smoke detection systems to waiting rooms open to corridors in all Group I occupancies, including Group I-2 hospitals. The provisions require only that a manual fire alarm system and an automatic fire detection system be installed in all occupancies classified as Group I.

[F] 907.2.6.2 Group I-2. Corridors in nursing homes (both intermediate care and skilled nursing facilities), detoxification facilities and spaces permitted to be open to the corridors by Section 407.2 shall be equipped with an automatic fire detection system. Hospitals shall be equipped with smoke detection as required in Section 407.2.

Exceptions:

1. Corridor smoke detection is not required in smoke compartments that contain patient sleeping units where patient sleeping units are provided with smoke detectors that comply with UL 268. Such detectors shall provide a visual display on the corridor side of each patient sleeping unit and an audible and visual alarm at the nursing station attending each unit.
2. Corridor smoke detection is not required in smoke compartments that contain patient sleeping units

where patient sleeping unit doors are equipped with automatic door-closing devices with integral smoke detectors on the unit sides installed in accordance with their listing, provided that the integral detectors perform the required alerting function.

Q: Is an electrically supervised, smoke detection system required throughout a Group I-2 occupancy (hospital)?

A: No. Section 907.2.6.2 contains no requirements specific to a Group I-2 hospital. The provisions do require an automatic fire detection system in corridors and all spaces open to corridors in Group I-2 nursing homes and detoxification facilities. Section 907.2.6 requires installation of electrically supervised, automatic smoke detection systems only in waiting areas that are open to corridors in all Group I occupancies.

[F] 907.2.7 Group M. A manual fire alarm system shall be installed in Group M occupancies having an occupant load of 500 or more persons or more than 100 persons above or below the lowest level of exit discharge. The initiation of a signal from a manual fire alarm box shall initiate alarm notification appliances as required by Section 907.9.

Exceptions:

1. Covered mall buildings complying with Section 402.
2. Manual fire alarm boxes are not required where the building is equipped throughout with an automatic sprinkler system and the alarm notification appliances will automatically activate upon sprinkler water flow.

Q: Given a building on grade without a basement, see Attachment No. 1 Multi-Story Building. Is the second story of this building the first story above the lowest level of exit discharge?

A: No. When determining stories above the lowest level of exit discharge, a level, or floor level, is not a story. A level is the horizontal plane that is part of a story, not the entire story height. A story is the vertical space between the upper surface of one floor level and the upper surface of the floor level next above or below.

The level of exit discharge as shown in attached drawing of a Multi-Story Building is at elevation 0.0 feet, the first floor level of the building is also at elevation 0.0 feet; therefore, the level of exit discharge and the first floor level of the building are at the same elevation. The first story of the building begins at elevation 0.0 feet (first floor) and extends to elevation 10.0 feet (second floor). The first story of the building is the first story above the level of exit discharge.

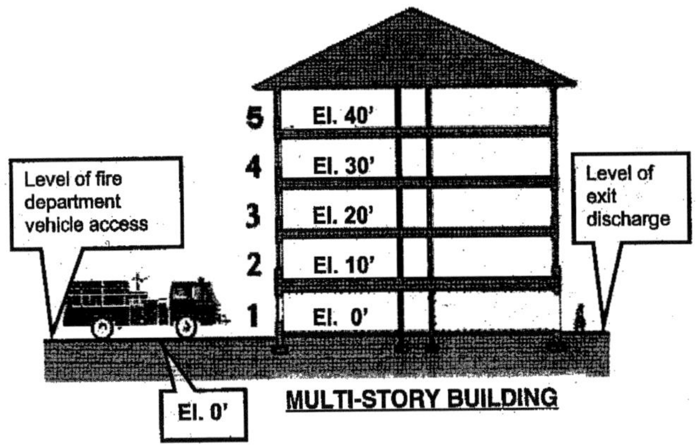

Attachment No. 1

Q: Is it the intent for the provisions of Section 907.2.7 to require a manual fire alarm system in a single-story Group M occupancy with an occupant load of less than 500 persons?

A: No. The provisions of Section 907.2.7 do not require a manual fire alarm system in a Group M occupancy where the first story is the lowest level of exit discharge and the occupants having access to exit and exit discharge without being required to travel up or down an exit stairway does not exceed 100 persons. A manual fire alarm system is required when a Group M occupancy has 500 or more occupants, regardless of where they are located within the building.

ICC Committee Interpretation 44-03, *International Fire Code,* Section 907.2.7, issued May 11, 2004 addresses how to identify the location of stories relative to the level of exit discharge, and is not intended to address or clarify the specific scoping requirements for manual fire alarm systems in Group M occupancies.

This provision for a manual fire alarm system has its origin in the *Standard Building Code*, and used the term "above or below the street floor level," which was not defined. When the provision for a manual fire alarm system was brought forward into the working draft of the 2000 *International Building Code,* the term "above or below the level of exit discharge" was adopted; the original term "above or below the street floor level," used in the *Standard Building Code* was dropped. It was not the intent of the provisions in the *Standard Building Code* to require a manual fire alarm system when the minimum occupant load is located at or on the street floor level or on the "level of exit discharge."

In addition, "level of exit discharge" became a defined term in the *International Building Code* and no correlation was made to address the difference in the meaning and use of the two terms. As defined in the *International Building Code*, the "Level of exit discharge" is: "The horizontal plane located at the point at which an exit terminates and an exit discharge begins."

[F] 907.2.9 Group R-2. A manual fire alarm system shall be installed in Group R-2 occupancies where:

1. Any dwelling unit or sleeping unit is located three or more stories above the lowest level of exit discharge;
2. Any dwelling unit or sleeping unit is located more than one story below the highest level of exit discharge of exits serving the dwelling unit or sleeping unit; or
3. The building contains more than 16 dwelling units or sleeping units.

Exceptions:

1. A fire alarm system is not required in buildings not more than two stories in height where all dwelling units or sleeping units and contiguous attic and crawl spaces are separated from each other and public or common areas by at least 1-hour fire partitions and each dwelling unit or sleeping unit has an exit directly to a public way, exit court or yard.
2. Manual fire alarm boxes are not required throughout the building when the following conditions are met:
   2.1. The building is equipped throughout with an automatic sprinkler system in accordance with Section 903.3.1.1 or Section 903.3.1.2; and
   2.2. The notification appliances will activate upon sprinkler flow.
3. A fire alarm system is not required in buildings that do not have interior corridors serving dwelling units and are protected by an approved automatic sprinkler system installed in accordance with Section 903.3.1.1 or 903.3.1.2, provided that dwelling units either have a means of egress door opening directly to an exterior exit access that leads directly to the exits or are served by open-ended corridors designed in accordance with Section 1023.6, Exception 4.

Q: Given a building on grade without a basement, see Attachment No. 1 Multi-Story Building. Is the second story of this building the first story above the lowest level of exit discharge?

A: No. When determining stories above the lowest level of exit discharge, a level, or floor level, is not a story. A level is the horizontal plane that is part of a story, not the entire story height. A story is the vertical space between the upper surface of one floor level and the upper surface of the floor level next above or below.

The level of exit discharge as shown in attached drawing of a Multi-Story Building is at elevation 0.0 feet, the first floor level of the building is also at elevation 0.0 feet ; therefore, the level of exit discharge and the first floor level of the building are at the same elevation. The first story of the building begins at elevation 0.0 feet (first floor) and extends to elevation 10.0 feet (second floor). The first story of the building is the first story above the level of exit discharge.

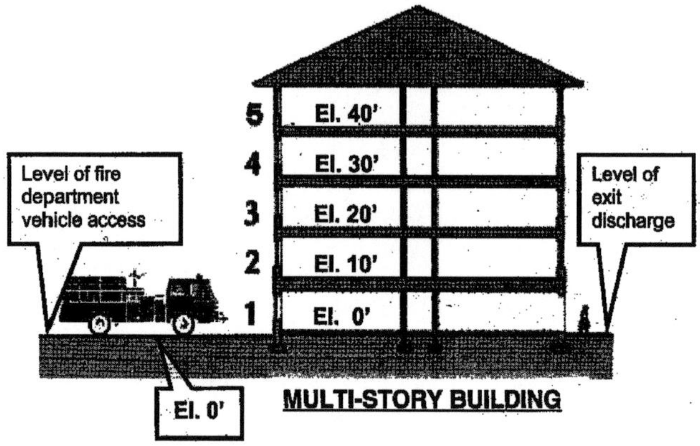

Attachment No. 1

[F] 907.2.10.1.2 Groups R-2, R-3, R-4 and I-1. Single- or multiple-station smoke alarms shall be installed and maintained in Groups R-2, R-3, R-4 and I-1, regardless of occupant load at all of the following locations:

1. On the ceiling or wall outside of each separate sleeping area in the immediate vicinity of bedrooms.
2. In each room used for sleeping purposes.
3. In each story within a dwelling unit, including basements but not including crawl spaces and uninhabitable attics. In dwellings or dwelling units with split levels and without an intervening door between the adjacent levels, a smoke alarm installed on the upper level shall suffice for the adjacent lower level provided that the lower level is less than one full story below the upper level.

Q: Do the provisions of Section 907.2.10.1.2 in the *International Building Code* require smoke alarms to be installed in all new one- and two-family dwellings?

A: No. The smoke alarm provisions of the *International Fire Code* are only applicable to one- and two-family dwelling buildings constructed under the *International Building Code* and classified in Group R by that code. They do not apply to detached one- and two-family dwellings and townhouses not regulated as Group R occupancies and constructed under the *International Residential Code.*

The ICC Board of Directors, in originally approving the development of the *International Residential Code*, indicated that it is to be a stand-alone code for the construction of detached one- and two-family dwellings and multiple single-family dwellings (townhouses) not more than three stories in height with a separate means of egress. The provisions for new construction that affect these type buildings are to be regulated exclusively by the *International Residential Code* and are not to be regulated by any other *International Code*. The exception to Section 101.2 of the *International Building Code* stipulates that detached one- and two-family dwellings and multiple single-family dwellings (townhouses) that are not more than three stories above grade plane in height shall comply with the *International Residential Code*. Likewise, Section R101.2, of the *International Residential Code* states that the provisions of the *International Residential Code* shall apply to the construction of detached one- and two-family dwellings and multiple single-family dwellings (townhouses) not more than three stories above grade in height with a separate means of egress and their accessory structures.Therefore, the construction of detached one- and two-family dwellings and multiple single-family dwellings (townhouses) within the scope of the *International Residential Code* is not required to comply with the provisions of either the *International Building Code* or the *International Fire Code*.

Where a dwelling or townhouse exceeds the allowable height in stories, does not provide individual egress for each dwelling unit, or does not conform to the prescriptive provisions of the *International Residential Code*, the structures are then beyond the scope of that code and its provisions cannot be applied. The building must then meet the provisions of the *International Building Code* for Group R-3 as well as the applicable provisions of the *International Fire Code*.

Single- and multiple-station smoke alarm provisions for buildings not regulated in Occupancy Group R constructed under the *International Residential Code* are contained in Section R313 of that code and require that smoke alarms be installed in both new construction and buildings undergoing alterations, repairs or additions.

[F] 907.2.12.2 Emergency voice/alarm communication system. The operation of any automatic fire detector, sprinkler water-flow device or manual fire alarm box shall automatically sound an alert tone followed by voice instructions giving approved information and directions for a general or staged evacuation on a minimum of the alarming floor, the floor above and the floor below in accordance with the building's fire safety and evacuation plans required by Section 404 of the *International Fire Code*. Speakers shall be provided throughout the building by paging zones. As a minimum, paging zones shall be provided as follows:

1. Elevator groups.
2. Exit stairways.
3. Each floor.
4. Areas of refuge as defined in Section 1002.1.

   Exception: In Group I-1 and I-2 occupancies, the alarm shall sound in a constantly attended area and a general occupant notification shall be broadcast over the overhead page.

Q: Section 907.2.12.2 requires that the voice/alarm signaling system have the capability to sound automatically an alert signal to all occupants located within the building on a general or selective basis to specified terminal areas. Does the code require that the voice/alarm signaling system selectively identify and send messages to certain areas of the building?

A: No. The intent of these requirements is to provide the capability to send out selective messages to individual terminal areas of the building. This is not intended to prohibit the same message from going out to all terminal areas. However, the system must also contain controls within the fire command station to allow manual transmission of an evacuation signal and voice instructions on a selective and all-call basis to only the specified terminal areas.

Q: Section 907.2.12.2 identifies four different terminal areas requiring a voice/alarm signaling system to be established on a general or selective basis. Does this mean that there shall be a speaker in each of these areas to carry the voice communication for alarm systems?

A: Yes. The four areas identified are considered terminal areas. Each one is expected to have its own speakers, enabling selective messages to be provided.

[F] 907.9.1 Visible alarms. Visible alarm notification appliances shall be provided in accordance with Sections 907.9.1.1 through 907.9.1.4.

Exceptions:

1. Visible alarm notification appliances are not required in alterations, except where an existing fire alarm system is upgraded or replaced, or a new fire alarm system is installed.
2. Visible alarm notification appliances shall not be required in exits as defined in Section 1002.1.

Q: Do the provisions of Sections 907.9.1 and 907.9.2 allow the option of providing either visible or audible alarm-notification appliances?

A: No. There is no option of providing either visible or audible alarms. If the building is a type that requires visible alarms, the devices must be heard as well. However, the code acknowledges the marginal value of audible alarms in certain areas where the average ambient noise is greater than 105 dBa. In these areas, only visible alarm-notification appliances in accordance with NFPA 72 are required.

[F] 907.9.2 Audible alarms. Audible alarm notification appliances shall be provided and shall sound a distinctive sound that is not to be used for any purpose other than that of a fire alarm. The audible alarm notification appliances shall provide a sound pressure level of 15 decibels (dBA) above the average ambient sound level or 5 dBA above the maximum sound level having a duration of at least 60 seconds, whichever is greater, in every occupied space within the building. The minimum sound pressure levels shall be: 70 dBA in occupancies in Group R and Group I-1; 90 dBA in mechanical equipment rooms; and 60 dBA in other occupancies. The maximum sound pressure level for audible alarm notification appliances shall be 120 dBA at the minimum hearing distance from the audible appliance. Where the average ambient noise is greater than 105 dBA, visible alarm notification appliances shall be provided in accordance with NFPA 72 and audible alarm notification appliances shall not be required.

Exception: Visible alarm notification appliances shall be allowed in lieu of audible alarm notification appliances in critical care areas of Group I-2 occupancies.

Q: In an occupancy required to be equipped with a fire alarm system, is the audible alarm required to provide the minimum sound pressure level throughout bathrooms, mechanical equipment rooms and other such spaces of the building?

A: Yes. Section 907.9.2 requires that an audible alarm be heard in all occupied spaces of the building. The intent of the audible alarm is to warn occupants throughout the building of a fire condition. The alarm signal must be heard in rooms and spaces defined as "habitable" and "occupiable" as well as other spaces which are utilized by the occupants, such as bathrooms, storage rooms and mechanical rooms. Mechanical equipment rooms are specifically identified in this section due to the presence of equipment that, when operating, may necessitate an increase in the minimum sound pressure level based on higher ambient sound levels. Spaces that do not require audible warning include plenums, crawl spaces, attics and other spaces that are not intended, designed or arranged to be occupied or utilized by occupants. However, the code acknowledges the marginal value of audible alarms in certain areas where the average ambient noise is greater than 105 dBa. In these areas, only visible alarm notification appliances in accordance with NFPA 72 are required.

CHAPTER 10

MEANS OF EGRESS

SECTION 1002 DEFINITIONS

1002.1 Definitions. The following words and terms shall, for the purposes of this chapter and as used elsewhere in this code, have the meanings shown herein.

FLOOR AREA, GROSS. The floor area within the inside perimeter of the exterior walls of the building under consideration, exclusive of vent shafts and courts, without deduction for corridors, stairways, closets, the thickness of interior walls, columns or other features. The floor area of a building, or portion thereof, not provided with surrounding exterior walls shall be the usable area under the horizontal projection of the roof or floor above. The gross floor area shall not include shafts with no openings or interior courts.

Q: Does "gross floor area" apply to all areas under the roof, including eaves and overhangs?

A: No. The gross floor area is the area within the inside perimeter of the exterior walls exclusive of areas open and unobstructed to the sky.

MEANS OF EGRESS. A continuous and unobstructed path of vertical and horizontal egress travel from any occupied portion of a building or structure to a public way. A means of egress consists of three separate and distinct parts: the exit access, the exit and the exit discharge.

Q: Are exterior stairways that are located anywhere between a building and the public way considered a part of the means of egress from a building?

A: Yes. The path of travel from the exit discharge doors to a public way is considered exit discharge in accordance with the definitions of "Means of egress" and "Exit discharge" in Section 1002. Any stairway constituting a portion of this path of travel would be considered an exit discharge stairway.

SECTION 1003 GENERAL MEANS OF EGRESS

1003.1 Applicability. The general requirements specified in Sections 1003 through 1013 shall apply to all three elements of the means of egress system, in addition to those specific requirements for the exit access, the exit and the exit discharge detailed elsewhere in this chapter.

Q: Can the required capacity of each egress element from a room or space be determined by dividing the design occupant load of that space by the number of exits and exit access doorways that lead directly from the space?

A: Yes. Section 1003.1 requires the total capacity of all egress elements from a room or space to equal or exceed the design occupant load of the space as determined in accordance with Section 1004.1. Be aware that the design occupant load is the largest number of occupants derived by the application of Sections 1004.1.1 and also 1004.7.

An equal distribution of occupants between all egress elements leading out of a space reasonably approximates the use of multiple egress components from most occupancies should emergency evacuation be required. Note, however, that Section 1025.2 requires a more conservative egress distribution for assembly occupancies based on higher occupant load densities. Those elements that contribute to the required egress capacity from the space (e.g., exterior exit door, exit stairway or exit access doorway) must be identified with "Exit" signs as required by Section 1011.1 and comply with the other applicable provisions of Chapter 10.

Q: Is it mandatory for nonrequired doorways to contribute to the egress capacity of the room or space from which they lead?

A: No. The minimum required number of exits and exit access elements specified in Sections 1015.1 and 1019.1 provide an acceptable level of safety for the occupants. Although additional doorways are permitted, they are not specifically required to comply with all provisions of Chapter 10. Nonrequired doorways are to be constructed such that they serve their intended purpose without creating a hazard to the building occupants or general public. Accordingly, nonrequired doorways must not be identified as means of egress unless they fully comply with the applicable provisions of Chapter 10.

1003.4 Floor surface. Walking surfaces of the means of egress shall have a slip-resistant surface and be securely attached.

Q: A building contains exterior balconies, decks, porches and patios that are considered occupiable spaces as regulated by the code. Does Section 1003.4 require the walking surfaces of these building elements to be slip-resistant?

A: Yes. Because the pace of egress travel can become hurried during emergency situations, the probability of slipping on smooth or slick floor surfaces increases. To minimize the hazard, all floor surfaces in the means of egress are required to be slip resistant. As occupiable spaces, exterior balconies, decks, porches and patios are no different than interior building spaces, which are required to be provided with adequate access to the required exits. Accordingly, the occupants of the building's occupiable exterior elements must also have access to those same exits. The walking surfaces of those exterior elements are part of the means of egress, specifically exit access, and must have a slip-resistant surface. The code requires an acceptable minimum level of safety for all walking surfaces throughout the means of egress. The use of floor materials with highly polished, glazed, glossy or finely finished surfaces must be avoided.

SECTION 1004
OCCUPANT LOAD

1004.1 Design occupant load. In determining means of egress requirements, the number of occupants for whom means of egress facilities shall be provided shall be determined in accordance with this section. Where occupants from accessory areas egress through a primary space, the calculated occupant load for the primary space shall include the total occupant load of the primary space plus the number of occupants egressing through it from the accessory area.

Q: The assembly area in a Group A-2 night club provides for seating at tables and chairs (unconcentrated), and also contains a dance floor. Is the area of the dance floor included in determining the minimum occupant load?

A: Yes. The code does not address the issue of non-simultaneous occupancy. Although the patrons on the dance floor could also have a seat at a table, the code requires the "net floor area" as defined by the code to be included in the calculation of the occupant load. Therefore, an occupant load must be determined for the seating area as well as the dance floor.

1004.1.1 Areas without fixed seating. The number of occupants shall be computed at the rate of one occupant per unit of area as prescribed in Table 1004.1.1. For areas without fixed seating, the occupant load shall not be less than that number determined by dividing the floor area under consideration by the occupant per unit of area factor assigned to the occupancy as set forth in Table 1004.1.1. Where an intended use is not listed in Table 1004.1.1, the building official shall establish a use based on a listed use that most nearly resembles the intended use.

> **Exception:** Where approved by the building official, the actual number of occupants for whom each occupied space, floor or building is designed, although less than those determined by calculation, shall be permitted to be used in the determination of the design occupant load.

Q: Can a floor area be designed for a higher occupant load than that determined by Table 1004.1.1?

A: Yes. Section 1004.1 establishes that the minimum design occupant load for which exits must be provided is the largest number computed from Sections 1004.1 through 1004.1.1. Section 1004.2 permits a building to be used and occupied by a higher occupant load, provided that the building fully complies with the code based on the higher number of occupants. (Note: The maximum occupant load, in any case, is not permitted to exceed one occupant per 7 square feet of occupiable floor space in accordance with Section 1004.2.)

SECTION 1005
EGRESS WIDTH

1005.1 Minimum required egress width. The means of egress width shall not be less than required by this section. The total width of means of egress in inches (mm) shall not be less than the total occupant load served by the means of egress multiplied by the factors in Table 1005.1 and not less than specified elsewhere in this code. Multiple means of egress shall be sized such that the loss of any one means of egress shall not reduce the available capacity to less than 50 percent of the required capacity. The maximum capacity required from any story of a building shall be maintained to the termination of the means of egress.

> **Exception:** Means of egress complying with Section 1025.

Q: Is it the intent of Section 1005.1 to consider a pair of doors without a mullion as two separate openings for the purpose of determining the means of egress occupant capacity from Table 1005.1?

A: No. A pair of doors without a center mullion can be considered a single opening, provided that both doors are equipped with hardware complying with the requirements of Section 1008.1. The opening width to be used in computing the means of egress occupant capacity is the clear opening width with both doors open. (Note: One leaf must provide a minimum clear opening width of 32 inches in accordance with Section 1008.1.1 for those occupancies requiring a door opening to have a 32-inch clear width.)

SECTION 1007
ACCESSIBLE MEANS OF EGRESS

1007.2.1 Elevators required. In buildings where a required accessible floor is four or more stories above or below a level of exit discharge, at least one required accessible means of egress shall be an elevator complying with Section 1007.4.

Exceptions:

1. In buildings equipped throughout with an automatic sprinkler system installed in accordance with Section 903.3.1.1 or 903.3.1.2, the elevator shall not be required on floors provided with a horizontal exit and located at or above the level of exit discharge.
2. In buildings equipped throughout with an automatic sprinkler system installed in accordance with Section 903.3.1.1 or 903.3.1.2, the elevator shall not be required on floors provided with a ramp conforming to the provisions of Section 1010.

Q: Given a building on grade, without a basement, see Attachment No. 1 Multi-Story Building. Is the second story of this building the first story above the lowest level of exit discharge?

A: No. When determining stories above the lowest level of exit discharge, a level, or floor level, is not a story. A "level" is the horizontal plane that is part of a story, not the entire story height. A "story" is the vertical space between the upper surface of one floor level and the upper surface of the floor level next above or below.

The level of exit discharge as shown in attached drawing of a Multi-Story Building is at elevation 0.0 feet, the first floor level of the building is also at elevation 0.0 feet; therefore, the level of exit discharge and the first floor level of the building are at the same elevation. The first story of the building begins at elevation 0.0 feet (first floor) and extends to elevation 10.0 feet (second floor). The first story of the building is the first story above the level of exit discharge.

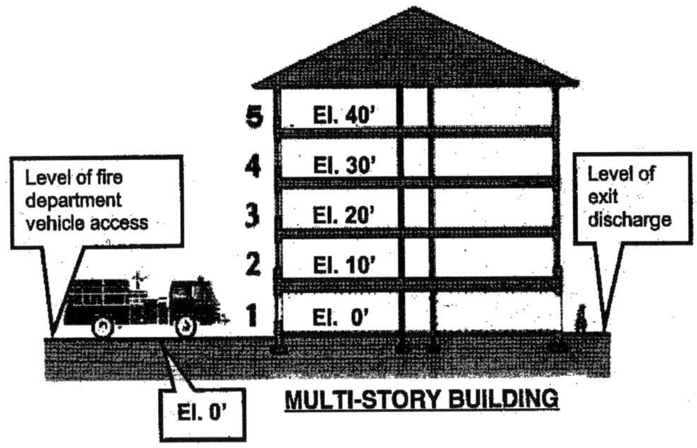

Attachment No. 1

SECTION 1008
DOORS, GATES AND TURNSTILES

1008.1.2 Door swing. Egress doors shall be side-hinged swinging.

Exceptions:

1. Private garages, office areas, factory and storage areas with an occupant load of 10 or less.
2. Group I-3 occupancies used as a place of detention.
3. Critical or intensive care patient rooms within suites of health care facilities.
4. Doors within or serving a single dwelling unit in Groups R-2 and R-3.
5. In other than Group H occupancies, revolving doors complying with Section 1008.1.3.1.
6. In other than Group H occupancies, horizontal sliding doors complying with Section 1008.1.3.3 are permitted in a means of egress.
7. Power-operated doors in accordance with Section 1008.1.3.2.
8. Doors serving a bathroom within an individual sleeping unit in Group R-1.

Doors shall swing in the direction of egress travel where serving an occupant load of 50 or more persons or a Group H occupancy.

The opening force for interior side-swinging doors without closers shall not exceed a 5-pound (22 N) force. For other side-swinging, sliding and folding doors, the door latch shall release when subjected to a 15-pound (67 N) force. The door shall be set in motion when subjected to a 30-pound (133 N) force. The door shall swing to a full-open position when subjected to a 15-pound (67 N) force. Forces shall be applied to the latch side.

Q: A room with an occupancy in Group A, B, E, F or M with an occupant load of more than 49 individuals is required by Table 1015.1 to have two means of egress doors. Are these doors required to swing in the direction of egress travel?

A: Yes. Section 1008.1.2 requires all doors serving an occupant load of 50 or more or a high-hazard occupancy to swing in the direction of egress travel. The occupant load of 50 or more relates to the occupant load of the room and not to the exiting capacity of the doors.

1008.1.3.2 Power-operated doors. Where means of egress doors are operated by power, such as doors with a photoelectric-actuated mechanism to open the door upon the approach of a person, or doors with power-assisted manual operation, the design shall be such that in the event of power failure, the door is capable of being opened manually to permit means of egress travel or closed where necessary to safeguard means of egress. The forces required to open these doors manually shall not exceed those specified in Section 1008.1.2, except that the force to set the door in motion shall not exceed 50 pounds (220 N). The door shall be capable of swinging from any position to the full width of the opening in which such door is installed when a force is applied to the door on the side from which egress is made. Full-power-operated doors shall comply with BHMA A156.10. Power-assisted and low-energy doors shall comply with BHMA A156.19.

Exceptions:

1. Occupancies in Group I-3.
2. Horizontal sliding doors complying with Section 1008.1.3.3.
3. For a biparting door in the emergency breakout mode, a door leaf located within a multiple-leaf opening shall be exempt from the minimum 32-inch (813 mm) single-leaf requirement of Section 1008.1.1, provided a minimum 32-inch (813 mm) clear opening is provided when the two biparting leaves meeting in the center are broken out.

Q: Can the opening that results from the emergency swing of adjacent panels of power-operated sliding or swinging doors be used to meet the minimum required clear width?

A: Yes. Section 1008.1.3.2 deals with the maximum force to be applied to power-operated doors that will permit them to be considered as an egress element. These forces must open the normally power-operated panels in a swing mode during a loss of power.

Where Section 1008.1.1 requires a minimum clear width of 32 inches and a door opening includes two door leaves without a mullion, one leaf shall provide the clear opening width of 32 inches, regardless of the door operation.

1008.1.3.3 Horizontal sliding doors. In other than Group H occupancies, horizontal sliding doors permitted to be a component of a means of egress in accordance with Exception 5 to Section 1008.1.2 shall comply with all of the following criteria:

1. The doors shall be power operated and shall be capable of being operated manually in the event of power failure.
2. The doors shall be openable by a simple method from both sides without special knowledge or effort.
3. The force required to operate the door shall not exceed 30 pounds (133 N) to set the door in motion and 15 pounds (67 N) to close the door or open it to the minimum required width.
4. The door shall be openable with a force not to exceed 15 pounds (67 N) when a force of 250 pounds (1100 N) is applied perpendicular to the door adjacent to the operating device.
5. The door assembly shall comply with the applicable fire protection rating and, where rated, shall be self-closing or automatic-closing by smoke detection, shall be installed in accordance with NFPA 80 and shall comply with Section 715.
6. The door assembly shall have an integrated standby power supply.
7. The door assembly power supply shall be electrically supervised.
8. The door shall open to the minimum required width within 10 seconds after activation of the operating device.

Q: Does Section 1008.3 apply to security gates or grilles installed on the inside of means of egress doorways and that swing against the direction of egress travel if the gates are left open during occupancy?

A: No. Such a security gate can be dealt with under Section 1008.1.3.5.

Q: If gates are permitted, can they be equipped with either double-cylinder deadbolts requiring a key from both sides, or padlock and hasp?

A: No. A double-cylinder deadbolt or other type of key locking device is prohibited on egress doors in accordance with Section 1008.1.8 and would, therefore, not be permissible on a security gate in a doorway.

Q: Does Section 1008.1.3.3 permit the use of nonpower-operated horizontal sliding doors as a means of egress if the door, when pushed, will swing in the event of an emergency?

A: No. Except as provided for in the exceptions, Section 1008.1.3.3 requires all nonpower-operated egress doors to be of a side-swinging type. Furthermore, the door must swing in the direction of egress whenever serving an occupant load of 50 or more or a high-hazard occupancy. The normal operation of a horizontal sliding door is not of a side-swinging type. Although the door can swing open in the event of an emergency, the operation will likely not be known to the building occupants because the normal operation is to open horizontally.

Criteria for Section 1008.1.3.3 permit the use of horizontal sliding doors that comply with the conditions set forth therein. Power-operated sliding or swinging doors are required to comply with Section 1008.1.3.2 provisions.

1008.1.3.4 Access-controlled egress doors. The entrance doors in a means of egress in buildings with an occupancy in Group A, B, E, M, R-1 or R-2 and entrance doors to tenant spaces in occupancies in Groups A, B, E, M, R-1 and R-2 are permitted to be equipped with an approved entrance and egress access control system which shall be installed in accordance with all of the following criteria:

1. A sensor shall be provided on the egress side arranged to detect an occupant approaching the doors. The doors shall be arranged to unlock by a signal from or loss of power to the sensor.
2. Loss of power to that part of the access control system which locks the doors shall automatically unlock the doors.
3. The doors shall be arranged to unlock from a manual unlocking device located 40 inches to 48 inches (1016 mm to 1219 mm) vertically above the floor and within 5 feet (1524 mm) of the secured doors. Ready access shall be provided to the manual unlocking device and the device shall be clearly identified by a sign that reads "PUSH TO EXIT." When operated, the manual unlocking device shall result in direct interruption of power to the lock—independent of the access control system electronics—and the doors shall remain unlocked for a minimum of 30 seconds.
4. Activation of the building fire alarm system, if provided, shall automatically unlock the doors, and the doors shall remain unlocked until the fire alarm system has been reset.
5. Activation of the building automatic sprinkler or fire detection system, if provided, shall automatically unlock the doors. The doors shall remain unlocked until the fire alarm system has been reset.
6. Entrance doors in buildings with an occupancy in Group A, B, E or M shall not be secured from the egress side during periods that the building is open to the general public.

Q: Groups A, B, E and M occupancies, if criteria 1 through 5 are met and the access-controlled egress doors unlock on the egress side prior to an occupant reaching the doors and no subsequent securing devices are installed, does the access-control system need to be disabled during the hours the building is open to the public?

A: No, the access-controlled system at egress doors is permitted to remain operational at all times; however, Item 6 specifically requires that during the hours the building is open to the general public, egress doors equipped with a controlled-access system shall not be secured from the egress side. It is permissible to lock occupants out of the building, but occupants may not be locked in the building. This restriction applies to Group A, B, E and M occupancies.

Q: Are the requirements for access-controlled egress doors described in Section 1008.1.3.4 intended as an alternate method to panic hardware equipped with actuating switches?

A: No. The access-controlled egress requirements in Section 1008.1.3.4 are not permitted to serve as an alternative method for egress hardware required elsewhere in the code. The requirements in Section 1008.1.3.4 provide additional controls beyond the minimum requirements, which may be added to control access through an entrance door that also serves as an egress or exit door. Where panic hardware is installed and secured during periods when the building is open to the public, the panic hardware itself and the electronic actuating switches within the panic hardware must be operational for egress as required in Items 1 through 6.

1008.1.8 Door operations. Except as specifically permitted by this section egress doors shall be readily openable from the egress side without the use of a key or special knowledge or effort.

Q: Section 1008.1.8 states that all means of egress doors must be readily openable from the side from which egress is to be made without the use of a key, special knowledge or effort. Where there is a series of adjacent doors for convenience but only one of them is required to satisfy the means of egress width based on the occupant load, can all but one be provided with a lock requiring the use of a key? The remaining door is equipped with a latch, that is readily recognized as being latched, and its operation does not require special knowledge or effort.

A: Yes. If supplemental adjacent doors are provided in excess of those required to satisfy the means of egress width based on the occupant load, the additional doors are not required to comply with the requirements for egress doors, provided that a means of channeling the natural flow of traffic to the latched door through the use of railings, ropes or other equivalent means is employed. The required egress door must also be clearly marked in accordance with Section 1011.

1008.1.8.3 Locks and latches. Locks and latches shall be permitted to prevent operation of doors where any of the following exists:

1. Places of detention or restraint.
2. In buildings in occupancy Group A having an occupant load of 300 or less, Groups B, F, M and S, and in places of religious worship, the main exterior door or doors are permitted to be equipped with key-operated locking devices from the egress side provided:
   2.1. The locking device is readily distinguishable as locked,
   2.2. A readily visible durable sign is posted on the egress side on or adjacent to the door stating: THIS DOOR TO REMAIN UNLOCKED WHEN BUILDING IS OCCUPIED. The sign shall be in letters 1 inch (25 mm) high on a contrasting background,
   2.3. The use of the key-operated locking device is revokable by the building official for due cause.
3. Where egress doors are used in pairs, approved automatic flush bolts shall be permitted to be used, provided that the door leaf having the automatic flush bolts has no doorknob or surface-mounted hardware.
4. Doors from individual dwelling or sleeping units of Group R occupancies having an occupant load of 10 or less are permitted to be equipped with a night latch, dead bolt or security chain, provided such devices are openable from the inside without the use of a key or tool.

Q: The only means of access to and egress from an exterior balcony that serves a dwelling unit requires an occupant to travel through the dwelling unit. Is the door separating the balcony from the interior of the dwelling unit permitted to be locked from the interior?

A: Yes. Exception 4 to Section 1008.1.8.3 allows egress doors from an individual dwelling unit with an occupant load of 10 or less to be equipped with a night latch, dead bolt or security chain to maintain the security of the occupants. Similarly, the egress door from the balcony is permitted to be equipped with any of the types of security measures identified in Exception 4 or hardware that is locked manually from the interior. Self-locking hardware that could accidentally and automatically trap an occupant on the balcony is prohibited.

Q: Are the provisions of Section 1008.1.8.3, Exception 2, of the *International Building Code* applicable to buildings with one exit as regulated in Section 1014.1?

A: Yes. The requirements of Section 1008.1.8.3, Exception 2, establish specific conditions under which a building's "main exterior" means of egress door may be equipped with locking hardware that requires the use of a key to operate but do not limit their application to any particular building configuration. Because the definition of "Means of egress door" does not include the descriptor "main exterior," it must assume the ordinarily accepted meaning (see Section 201.4) which, in this case, would be the principal means of egress door located in the exterior wall of a building and would likely also be considered the principal means of entrance. Because a building with one exit as regulated by Section 1015.1 has but a single exterior means of egress door, it must naturally be considered the "main" door to which the provisions of Section 1008.1.8.3, Exception 2, are applicable.

Q: The main exterior doors of a Group A occupancy with an occupant load of 300 or less is permitted by the provisions of the *International Building Code* Section 1008.1.8.3, Exception 2, to be equipped with a key-operated locking device that, in the unlocked position, will allow the door to be readily openable.

Is the main exterior egress door in a Group A occupancy with an occupant load of 100 or more required to be equipped with panic hardware or fire exit hardware, provisioned in Section 1008.1.9, in addition to the key-operated locking device permitted in Section 1008.1.8.3?

A: No. Panic hardware or fire exit hardware as described in Section 1008.1.9 is not required on the main exterior egress door of a Group A occupancy where the door is equipped with a single key-operated locking device that complies with all the provisions of Section 1008.1.8.3, Exception 2. A single key-operated locking device is permitted as an exception or alternative to the primary requirements for locking or latching of egress doors. The permissive requirements of Section 1008.1.8.3, Exception 2 are very specific in identifying when and under what conditions the exception may be applied. The main exterior door that is permitted to be equipped with a keyed lock is also required to be unlocked while the building is occupied. A single locking device in the unlocked position will allow the door to swing freely, and panic hardware or fire exit hardware is only required on a door that is held in the closed position by a latch or lock.

Q: With the exception of doors requiring panic hardware in accordance with Section 1008.1.9, would all required egress doors be subject to the provisions of Section 1008.1.8.3, independent of the number of occupants occupying the building or space?

A: Yes. The provisions of Section 1008.1.8.3 make no exceptions for the number of occupants involved. However, Section 1008.1.8.3 provides for certain exceptions.

The provisions of Section 1008.1.9 for panic hardware are applicable as indicated for the occupancies and occupant load stipulated.

1008.1.8.6 Delayed egress locks. Approved, listed, delayed egress locks shall be permitted to be installed on doors serving any occupancy except Group A, E and H occupancies in buildings that are equipped throughout with an automatic sprinkler system in accordance with Section 903.3.1.1 or an approved automatic smoke or heat detection system installed in accordance with Section 907, provided that the doors unlock in accordance with Items 1 through 6 below. A building occupant shall not be required to pass through more than one door equipped with a delayed egress lock before entering an exit.

1. The doors unlock upon actuation of the automatic sprinkler system or automatic fire detection system.
2. The doors unlock upon loss of power controlling the lock or lock mechanism.
3. The door locks shall have the capability of being unlocked by a signal from the fire command center.
4. The initiation of an irreversible process which will release the latch in not more than 15 seconds when a force of not more than 15 pounds (67 N) is applied for 1 second to the release device. Initiation of the irreversible process shall activate an audible signal in the vicinity of the door. Once the door lock has been released by the application of force to the releasing device, relocking shall be by manual means only.

   Exception: Where approved, a delay of not more than 30 seconds is permitted.
5. A sign shall be provided on the door located above and within 12 inches (305 mm) of the release device reading: PUSH UNTIL ALARM SOUNDS. DOOR CAN BE OPENED IN 15 [30] SECONDS.
6. Emergency lighting shall be provided at the door.

Q: Do the requirements in Section 1008.1.8.6, Item 4 require both labeled fire doors and nonlabeled doors to be equipped with latching devices?

A: Yes. All the provisions in Section 1008.1.8.6 shall apply to any door where a delayed egress lock is used. The "latch" referenced in Item 4 is a requirement of the delayed egress locking mechanism (see Item 2), and is required to be installed as part of the locking system on both labeled fire door assemblies and nonrated doors.

A delayed egress locking system, as defined in Section 1008.1.8.6, is a nonmandatory locking system that is permitted to be used on egress doors; however, it must be used in conjunction with and in addition to locks and latches required elsewhere in the code. The delayed egress lock provides an approved means of restraining/releasing an egress door that is otherwise required to remain operable from the egress side.

1008.1.9 Panic and fire exit hardware. Where panic and fire exit hardware is installed, it shall comply with the following:

1. The actuating portion of the releasing device shall extend at least one-half of the door leaf width.
2. The maximum unlatching force shall not exceed 15 pounds (67 N).

Each door in a means of egress from a Group A or E occupancy having an occupant load of 50 or more and any Group H occupancy shall not be provided with a latch or lock unless it is panic hardware or fire exit hardware.

Exception: A main exit of a Group A occupancy in compliance with Section 1008.1.8.3, Item 2.

Electrical rooms with equipment rated 1,200 amperes or more and over 6 feet (1829 mm) wide that contain overcurrent devices, switching devices or control devices with exit access doors must be equipped with panic hardware and doors must swing in the direction of egress.

If balanced doors are used and panic hardware is required, the panic hardware shall be the push-pad type and the pad shall not extend more then one-half the width of the door measured from the latch side.

Q: Can a thumb-turn cylinder lock or head and floor bolts be installed on a means of egress door required to be equipped with panic hardware in accordance with Section 1008.1.9?

A: No. The function of panic hardware is to make the door free swinging upon its operation. The proposed hardware would eliminate the opportunity for emergency egress in a panic situation. Note: Section 1008.1.8.4 prohibits manually operated flush or surface bolts.

Q: Does Section 1008.1.9 require the installation of panic hardware for means of egress doors equipped with locking devices?

A: Yes. Section 1008.1.9 requires panic hardware to be provided on egress doors that serve high occupant load densities in order to ensure prompt opening of the door in an emergency. The panic hardware provisions of Section 1008.1.9 apply to required egress doors that have all of the following characteristics:

1. Are equipped with latching or locking devices;
2. Serve an occupant load of 50 or more; and
3. Serve an assembly or educational occupancy.

Q: Is a door that is required to be equipped with panic hardware permitted to be equipped with a deadbolt lock that operates independently of the panic hardware?

A: No. Section 1008.1.8 requires all egress doors to be readily openable from the side from which egress is made without the use of a key, special knowledge or special effort. If the door is provided with a locking device that is not capable of being unlocked by panic hardware, that door would clearly be unavailable for egress when the locking device is engaged. Due to the uncertainty that the locking device will be unlocked during all times of occupancy, panic hardware must release all latching and locking devices on the required egress door.

SECTION 1009 STAIRWAYS

1009.2 Headroom. Stairways shall have a minimum headroom clearance of 80 inches (2032 mm) measured vertically from a line connecting the edge of the nosings. Such headroom shall be continuous above the stairway to the point where the line intersects the landing below, one tread depth beyond the bottom riser. The minimum clearance shall be maintained the full width of the stairway and landing.

Exception: Spiral stairways complying with Section 1009.8 are permitted a 78-inch (1981 mm) headroom clearance.

Q: How is the minimum required headroom intended to be measured within all parts of a stairway in accordance with Section 1009.2?

A: The intent of the *International Building Code* regarding stairway headroom can best be described by considering a sloped reference line connecting the tread nosings. The minimum required headroom is represented by a sloped line parallel to and located 6 feet 8 inches vertically above the reference line for conventional stairways and 6 feet 6 inches vertically for spiral stairways.

1009.4 Stairway landings. There shall be a floor or landing at the top and bottom of each stairway. The width of landings shall not be less than the width of stairways they serve. Every landing shall have a minimum dimension measured in the direction of travel equal to the width of the stairway. Such dimension need not exceed 48 inches (1219 mm) where the stairway has a straight run.

Exceptions:

1. Aisle stairs complying with Section 1025.
2. Doors opening onto a landing shall not reduce the landing to less than one-half the required width. When fully open, the door shall not project more than 7 inches (178 mm) into a landing.

Q: Does the *International Building Code* permit a stairway landing to be shaped in a half circle rather than the typical rectangular-shaped area?

A: Yes. Section 1009.4 requires that the least dimension of a landing shall not be less than the required width of a stairway. If the landing provides an area described by an arc with a radius equal to the required stairway width, the minimum required space necessary for means of egress will be available subject to the restrictions of Section 1009.4.

SECTION 1011 EXIT SIGNS

1011.2 Illumination. Exit signs shall be internally or externally illuminated.

Exception: Tactile signs required by Section 1011.3 need not be provided with illumination.

Q: Is it the intent of Section 1011.1 that all exit signs, including supplemental directional signs, be illuminated?

A: Yes. Section 1011.2 clearly intends that all exit signs be illuminated. The *International Building Code* requires that "all required means of egress" be indicated with approved illuminated exit signs. This encompasses both the exit access and the exit. The purpose of the supplemental directional exit sign is to direct building occupants to the exit if it is not readily visible from the exit access. Therefore, it is just as important for directional signs to be illuminated as it is for the exit sign itself to be illuminated.

SECTION 1012 HANDRAILS

1012.4 Continuity. Handrail-gripping surfaces shall be continuous, without interruption by newel posts or other obstructions.

Exceptions:

1. Handrails within dwelling units are permitted to be interrupted by a newel post at a stair landing.
2. Within a dwelling unit, the use of a volute, turnout or starting easing is allowed on the lowest tread.
3. Handrail brackets or balusters attached to the bottom surface of the handrail that do not project horizontally beyond the sides of the handrail within 1.5 inches (38 mm) of the bottom of the handrail shall not be considered to be obstructions and provided further that for each 0.5 inch (13 mm) of additional handrail perimeter dimension above 4 inches (102 mm), the vertical clearance dimension of 1.5 inches (38 mm) shall be permitted to be reduced by 0.125 inch (3 mm).

Q: Are stairway handrails required to be constructed without obstructions, such as intermediate supports and posts?

A: Yes. Section 1012.4 requires continuous handrails to facilitate a reasonable degree of safety and assistance whenever ascending or descending the stairway. Obstructions that would significantly impair the ready use of the handrail gripping surfaces or that by configuration may cause injury through sharp or abrasive elements to individuals using the handrails are specifically prohibited.

Section 1002 defines "Stairway" as including the "necessary landings and platforms connecting them." Section 1012.4 states that stairways shall have continuous handrails on both sides.

Q: Are handrails required to extend around landings?

A: No. The code does not intend for handrails to be continuous around landings. Handrails are required to provide a horizontal or sloping rail for the occupant to grasp while negotiating the treads and risers. Section 1012.5 provides specific direction for handrail detailing at each landing.

SECTION 1013 GUARDS

1013.1 Where required. Guards shall be located along open-sided walking surfaces, mezzanines, industrial equipment platforms, stairways, ramps and landings that are located more than 30 inches (762 mm) above the floor or grade below. Guards shall be adequate in strength and attachment in accordance with Section 1607.7. Where glass is used to provide a guard or as a portion of the guard system, the guard shall also comply with Section 2407. Guards shall also be located along glazed sides of stairways, ramps and landings that are located more than 30 inches (762 mm) above the floor or grade below where the glazing provided does not meet the strength and attachment requirements in Section 1607.7.

Exception: Guards are not required for the following locations:

1. On the loading side of loading docks or piers.
2. On the audience side of stages and raised platforms, including steps leading up to the stage and raised platforms.
3. On raised stage and platform floor areas, such as runways, ramps and side stages used for entertainment or presentations.
4. At vertical openings in the performance area of stages and platforms.
5. At elevated walking surfaces appurtenant to stages and platforms for access to and utilization of special lighting or equipment.
6. Along vehicle service pits not accessible to the public.
7. In assembly seating where guards in accordance with Section 1025.14 are permitted and provided.

Q: Is it the intent of Section 1013.1 to regulate an open-sided walking surface that is not a required element of a means of egress?

A: No. Section 1013.1 only applies to areas that are a required element of a means of egress. Most open-sided walking surfaces, mezzanines and landings are likely to be an element of the exit access, the exit or the exit discharge. If, however, there is an open-sided walking surface that is not an element of a means of egress, it can be evaluated by the code official as a potentially unsafe condition in accordance with Section 115.

SECTION 1015 EXIT AND EXIT ACCESS DOORWAYS

1015.1 Exit or exit access doorways required. Two exits or exit access doorways from any space shall be provided where one of the following conditions exists:

1. The occupant load of the space exceeds the values in Table 1015.1.
2. The common path of egress travel exceeds the limitations of Section 1014.3.
3. Where required by Sections 1015.3, 1015.4 and 1015.5.

Exception: Group I-2 occupancies shall comply with Section 1014.2.2.

TABLE 1015.1
SPACES WITH ONE MEANS OF EGRESS

OCCUPANCY	MAXIMUM OCCUPANT LOAD
A, B, E[a], F, M, U	49
H-1, H-2, H-3	3
H-4, H-5, I-1, I-3, I-4, R	10
S	29

a. Daycare maximum occupant load is 10.

Q: Do the maximum limitations in Table 1015.1 for buildings with one exit apply only to the portion of the building served by the single exit? For example, a building has two or more areas; each of which individually meets the limitations set forth in Table 1015.1 and has access to one exit. These areas are not separated from each other by fire walls.

A: No. The intent of Table 1015.1 is to limit the overall building size and occupant load in which occupants have access to only one exit. The building described exceeds, as a whole, the limitations of Table 1015.1 and provides all occupants with access to only one exit. In order for this arrangement to comply with Section 1015.1, each area must be separated from all others by a fire wall so that each area can be considered a separate building.

1015.2.1 Two exits or exit access doorways. Where two exits or exit access doorways are required from any portion of the exit access, the exit doors or exit access doorways shall be placed a distance apart equal to not less than one-half of the length of the maximum overall diagonal dimension of the building or area to be served measured in a straight line between exit doors or exit access doorways. Interlocking or scissor stairs shall be counted as one exit stairway.

Exceptions:

1. Where exit enclosures are provided as a portion of the required exit and are interconnected by a 1-hour fire-resistance-rated corridor conforming to the requirements of Section 1017, the required exit separation shall be measured along the shortest direct line of travel within the corridor.
2. Where a building is equipped throughout with an automatic sprinkler system in accordance with Section 903.3.1.1 or 903.3.1.2, the separation distance of the exit doors or exit access doorways shall not be less than one-third of the length of the maximum overall diagonal dimension of the area served.

Q: Does the *International Building Code* permit the use of scissor stairs as exits?

A: Yes. The use of scissor type stairs is not prohibited in Chapter 10. However, scissor stairs shall be counted as one exit stairway.

SECTION 1016
EXIT ACCESS TRAVEL DISTANCE

1016.1 Travel distance limitations. Exits shall be so located on each story such that the maximum length of exit access travel, measured from the most remote point within a story to the entrance to an exit along the natural and unobstructed path of egress travel, shall not exceed the distances given in Table 1016.1.

Where the path of exit access includes unenclosed stairways or ramps within the exit access or includes unenclosed exit ramps or stairways as permitted in Section 1020.1, the distance of travel on such means of egress components shall also be included in the travel distance measurement. The measurement along stairways shall be made on a plane parallel and tangent to the stair tread nosings in the center of the stairway.

Exceptions:

1. Travel distance in open parking garages is permitted to be measured to the closest riser of open stairs.
2. In outdoor facilities with open exit access components and open exterior stairs or ramps, travel distance is permitted to be measured to the closest riser of a stair or the closest slope of the ramp.
3. Where an exit stair is permitted to be unenclosed in accordance with Exception 8 or 9 of Section 1020.1, the travel distance shall be measured from the most remote point within a building to an exit discharge.

TABLE 1016.1
EXIT ACCESS TRAVEL DISTANCE[a]

OCCUPANCY	WITHOUT SPRINKLER SYSTEM (feet)	WITH SPRINKLER SYSTEM (feet)
A, E, F-1, I-1, M, R, S-1	200	250[b]
B	200	300[c]
F-2, S-2, U	300	400[b]
H-1	Not Permitted	75[c]
H-2	Not Permitted	100[c]
H-3	Not Permitted	150[c]
H-4	Not Permitted	175[c]
H-5	Not Permitted	200[c]
I-2, I-3, I-4	150	200[c]

For SI: 1 foot = 304.8 mm.

a. See the following sections for modifications to exit access travel distance requirements:
Section 402: For the distance limitation in malls.
Section 404: For the distance limitation through an atrium space.
Section 1015.2: For increased limitation in Groups F-1 and S-1.
Section 1024.7: For increased limitation in assembly seating.
Section 1024.7: For increased limitation for assembly open-air seating.
Section 1018.2: For buildings with one exit.
Chapter 31: For the limitation in temporary structures.

b. Buildings equipped throughout with an automatic sprinkler system in accordance with Section 903.3.1.1 or 903.3.1.2. See Section 903 for occupancies where sprinkler systems according to Section 903.3.1.2 are permitted.

c. Buildings equipped throughout with an automatic sprinkler system in accordance with Section 903.3.1.1.

Q: An enclosed mezzanine, in compliance with Section 505.4, Exception 2 of the *International Building Code*, is provided with one exit stairway that discharges into a sprinklered lobby in accordance with Section 1024.1, and one unenclosed stairway. Are the maximum travel distance limitations of Section 1016 applicable to the means of egress from the enclosed mezzanine?

A: Yes. The maximum exit access travel distance identified in Table 1016.1 is measured from each and every occupiable point within the building (including the mezzanine level) to an exit. Although the occupants may be required by Section 1019.1 to have access to a second, third or fourth exit, only the distance to the nearest exit is regulated by the code. Therefore, the distance an occupant must travel to reach a second, third or fourth required exit is not limited.

The occupants of the mezzanine described above are provided with two means of egress. The first exit access path begins at the most remote point of the mezzanine, proceeds across the mezzanine to the unenclosed stairway, down the stairway and terminates at the exit serving the main floor. The second exit access path also begins at the most remote point of the mezzanine, proceeds across the mezzanine and terminates at the entrance of the enclosed exit stairway leading from the mezzanine level to the exit lobby. Traveling along one of these exit access paths, occupants of the mezzanine must reach the safety of an exit within the maximum travel distance identified in Section 1016. The distance traveled within the safety of exit elements, such as the enclosed exit stairway and the exit lobby, is not limited.

SECTION 1017 CORRIDORS

1017.1 Construction. Corridors shall be fire-resistance rated in accordance with Table 1017.1. The corridor walls required to be fire-resistance rated shall comply with Section 708 for fire partitions.

Exceptions:

1. A fire-resistance rating is not required for corridors in an occupancy in Group E where each room that is used for instruction has at least one door directly to the exterior and rooms for assembly purposes have at least one-half of the required means of egress doors opening directly to the exterior. Exterior doors specified in this exception are required to be at ground level.
2. A fire-resistance rating is not required for corridors contained within a dwelling or sleeping unit in an occupancy in Group R.
3. A fire-resistance rating is not required for corridors in open parking garages.
4. A fire-resistance rating is not required for corridors in an occupancy in Group B which is a space requiring only a single means of egress complying with Section 1015.1.

Q: Does Section 1017 require exit access corridors to be provided in buildings?

A: No. Section 1017 only provides requirements for exit access corridors whenever they are provided in a building. The *International Building Code* requires access to exits for every building, which can be accomplished by open spaces, aisles or corridors. A corridor is created when the path of travel to an exit is confined, and communication with and observation of adjacent floor areas are restricted.

Q: A floor plan, of which a portion is an open plan without corridors and an adjacent portion that is compartmented, contains a corridor that leads to the open plan area. Assuming the corridor enclosure is required to be fire-resistance rated, is it the intent of the code to require that an opening protective be installed at the point where the corridor enters the open area?

A: No. The relatively open areas adjacent to the corridor can be considered corridor extensions with rated walls and opening protectives enclosing both the narrow and wide portions of the corridor in a continuous fashion.

1017.2 Corridor width. The minimum corridor width shall be as determined in Section 1005.1, but not less than 44 inches (1118 mm).

Exceptions:

1. Twenty-four inches (610 mm)—For access to and utilization of electrical, mechanical or plumbing systems or equipment.
2. Thirty-six inches (914 mm)—With a required occupant capacity of less than 50.
3. Thirty-six inches (914 mm)—Within a dwelling unit.
4. Seventy-two inches (1829 mm)—In Group E with a corridor having a required capacity of 100 or more.
5. Seventy-two inches (1829 mm)—In corridors serving surgical Group I, health care centers for ambulatory patients receiving outpatient medical care, which causes the patient to be not capable of self-preservation.
6. Ninety-six inches (2438 mm)—In Group I-2 in areas where required for bed movement.

Q: Section 1017.2 indicates that the required width of corridors can be reduced to 36 inches where serving an occupant load of less than 50. In layouts where dead-end or cross aisles and corridors serve limited portions of a floor area, can such passageways, aisle accessways, aisles and corridors be analyzed individually in segments so that only the occupant load that would actually traverse the exit access element to reach a required exit be used to determine the required width?

A: Yes. Certain passageway, aisle and corridor arrangements can be readily evaluated to determine the actual occupant load served by such element. In these cases, it can often be recognized that a given segment does not actually serve the entire occupant load of the room. In such cases, only the portion of the occupant load that would traverse part of a passageway, aisle or corridor to reach an exit need be used to determine the required width.

Q: Does the 72-inch minimum required corridor width apply to every corridor in a Group E building having an occupant load greater than 100?

A: No. Item 4 of Section 1017.2 is intended to apply only to corridors, aisle accessways, aisles or passageways serving 100 or more occupants. Corridors serving 51 to 99 occupants are required to be at least 44 inches wide, whereas corridors serving 50 or fewer occupants are required to be at least 36 inches wide.

1017.3 Dead ends. Where more than one exit or exit access doorway is required, the exit access shall be arranged such that there are no dead ends in corridors more than 20 feet (6096 mm) in length.

Exceptions:

1. In occupancies in Group I-3 of Occupancy Condition 2, 3 or 4 (see Section 308.4), the dead end in a corridor shall not exceed 50 feet (15 240 mm).
2. In occupancies in Groups B and F where the building is equipped throughout with an automatic sprinkler system in accordance with Section 903.3.1.1, the length of dead-end corridors shall not exceed 50 feet (15 240 mm).
3. A dead-end corridor shall not be limited in length where the length of the dead-end corridor is less than 2.5 times the least width of the dead-end corridor.

Q: A door is installed in an exit access corridor serving two exits, and the door only swings in one direction. Does the installation of this door constitute a dead-end corridor for the occupants traveling against the door swing?

A: No. A dead-end corridor is established when occupants are required to retrace their steps to gain access to an exit. A doorway across a corridor does not block the path of travel in the corridor so that occupants have to retrace their steps to reach an exit. However, Section 1008.1.2 establishes when a door must swing in the direction of travel. If the doorway in question serves either an occupant load of 50 or more in each direction or a Group H occupancy, the door must be capable of swinging in both directions or two doors, each swinging in opposite directions, must be provided.

Q: When more than one exit is required by the code, Section 1019.1 requires occupants to have direct access to the exits in separate directions. Section 1017.3 appears to conflict with Section 1019.1 because a dead-end corridor does not provide access to the exits in separate directions. Is it the intent of these sections to permit a dead-end corridor to exist in a building required to have more than one exit?

A: Yes. The requirements of Section 1019.1 are intended to provide exits that are remote as practicable from each other in accordance with Section 1015.2.1. Dead-end exit access travel within the limits of Section 1017.3 is specifically intended to be permitted and represents the limit at which the exits would still be deemed to be as remote as practicable.

1017.4 Air movement in corridors. Corridors shall not serve as supply, return, exhaust, relief or ventilation air ducts.

Exceptions:

1. Use of a corridor as a source of makeup air for exhaust systems in rooms that open directly onto such corridors, including toilet rooms, bathrooms, dressing rooms, smoking lounges and janitor closets, shall be permitted, provided that each such corridor is directly supplied with outdoor air at a rate greater than the rate of makeup air taken from the corridor.
2. Where located within a dwelling unit, the use of corridors for conveying return air shall not be prohibited.
3. Where located within tenant spaces of 1,000 square feet (93 m^2) or less in area, utilization of corridors for conveying return air is permitted.

Q: Can a single HVAC duct system (supply, return, exhaust, relief, or ventilation) serve an exit access corridor and spaces open and/or adjacent to the exit access corridor where the air supplied into and returned out of the open and/or adjacent spaces is through the duct system and not through the exit access corridor?

A: Yes. Air supplied from a single HVAC duct system is permitted to serve an exit access corridor and spaces open and/or adjacent to the exit access corridor. Air supplied by the duct system into the spaces open and/or adjacent to the exit access corridor, either by the duct system or by other mechanical or nonmechanical means, is not permitted to be transferred into or through the exit access corridor as supply air, return air, make up air, ventilation or exhaust.

SECTION 1019 NUMBER OF EXITS AND CONTINUITY

1019.1 Minimum number of exits. All rooms and spaces within each story shall be provided with and have access to the minimum number of approved independent exits required by Table 1019.1 based on the occupant load of the story, except as modified in Section 1015.1 or 1019.2. For the purposes of this chapter, occupied roofs shall be provided with exits as required for stories. The required number of exits from any story, basement or individual space shall be maintained until arrival at grade or the public way.

Q: In a building required to have a minimum of two exits, can a stairway serve as a "pass-through" space for a corridor continuing beyond the stairway and leading to a second exit?

A: No. Section 1019.1 requires the exits to be independent. A means of egress arrangement requiring building occupants to enter or use one exit as the only means to gain access to a second exit does not provide independent exits.

Q: In a building in which two or more exits are required by Section 1019.1, can a room or tenant space provided with only one means of egress doorway, in accordance with Section 1015.1, discharge directly into one of the required exits?

A: No. Section 1019.1 states that every floor area shall be provided with the minimum number of approved independent exits. In the condition described above, the alternative exit is not independent of the exit into which the room or tenant space discharges. Access to the alternative exit depends on the primary exit remaining available for use during an emergency. If the primary exit is rendered unusable, the secondary exit will also be unavailable, effectively eliminating the independence of the alternative exit. The *International Building Code* would allow an egress door from a room or space to discharge directly into an exit, provided there is a second egress door from the room or space offering independent access to the other required exit(s).

This situation should not be confused with the issue presented by a lobby used as an exit element in accordance with Section 1024.1. In this case, the lobby serves as an exit element for the occupants located only on the upper floors. For occupants located on the lobby floor, however, the lobby serves as an exit access passageway and can be used to provide independent access to the required exits (exterior doors).

Q: Are the minimum number of exits regulated by Section 1019.1 required to be available to all floor area occupants?

A: Yes. The basic concepts regarding means of egress require that approved independent exits must be provided from every floor area of a building. The minimum number of exits serving every floor area is determined in accordance with Section 1019.1 based on the occupant load of the floor area in question. These exits can be any of the types permitted by the code. An unenclosed stairway, other than as allowed by the exceptions to Section 1020.1 and other than those serving mezzanines and penthouses, can only provide convenient or supplemental access to an adjacent floor, and must comply with the provisions of either Section 404 or 707.2. An egress arrangement would not be permitted wherein an occupant must travel to an adjacent story through such a stairway to gain access to a required exit.

1019.2 Buildings with one exit. Only one exit shall be required in buildings as described below:

1. Buildings described in Table 1019.2, provided that the building has not more than one level below the first story above grade plane.
2. Buildings of Group R-3 occupancy.
3. Single-level buildings with the occupied space at the level of exit discharge provided that the story or space complies with Section 1015.1 as a space with one means of egress.

TABLE 1019.2
BUILDINGS WITH ONE EXIT

OCCUPANCY	MAXIMUM HEIGHT OF BUILDING ABOVE GRADE PLANE	MAXIMUM OCCUPANTS (OR DWELLING UNITS) PER FLOOR AND TRAVEL DISTANCE
A, B[d], E[e], F, M, U	1 Story	49 occupants and 75 feet travel distance
H-2, H-3	1 Story	3 occupants and 25 feet travel distance
H-4, H-5, I, R	1 Story	10 occupants and 75 feet travel distance
S[a]	1 Story	29 occupants and 100 feet travel distance
B[b], F, M, S[a]	2 Stories	30 occupants and 75 feet travel distance
R-2	2 Stories[c]	4 dwelling units and 50 feet travel distance

For SI: 1 foot = 304.8 mm.

a. For the required number of exits for open parking structures, see Section 1019.1.1.

b. For the required number of exits for air traffic control towers, see Section 412.1.

c. Buildings classified as Group R-2 equipped throughout with an automatic sprinkler system in accordance with Section 903.3.1.1 or 903.3.1.2 and provided with emergency escape and rescue openings in accordance with Section 1026 shall have a maximum height of three stories above grade plane.

d. Buildings equipped throughout with an automatic sprinkler system in accordance with Section 903.3.1.1 with an occupancy in Group B shall have a maximum travel distance of 100 feet.

e. Day care maximum occupant load is 10.

Q: Does a basement story that is not a story above grade but contains an exterior "areaway" stairway leading from that story constitute the level of exit discharge?

A: No. Below-grade or partially below-grade stories are not at the level of exit discharge, just as a second story above grade is not at the level of exit discharge. In this instance, the exit discharge is located at the top of the exterior "areaway" stairway (at the grade elevation). Thus, a level below the level of exit discharge is required to have a minimum of two approved exits unless the building qualifies for a single exit.

SECTION 1020
VERTICAL EXIT ENCLOSURES

1020.1 Enclosures required. Interior exit stairways and interior exit ramps shall be enclosed with fire barriers constructed in accordance with Section 706 or horizontal assemblies constructed in accordance with Section 711, or both. Exit enclosures shall have a fire-resistance rating of not less than 2 hours where connecting four stories or more and not less than 1 hour where connecting less than four stories. The number of stories connected by the exit enclosure shall include any basements but not any mezzanines. An exit enclosure shall not be used for any purpose other than means of egress.

Exceptions:

1. In all occupancies, other than Group H and I occupancies, a stairway is not required to be enclosed when the stairway serves an occupant load of less than 10 and the stairway complies with either Item 1.1 or 1.2. In all cases, the maximum number of connecting open stories shall not exceed two.
   1.1. The stairway is open to not more than one story above the story at the level of exit discharge; or
   1.2. The stairway is open to not more than one story below the story at the level of exit discharge.
2. Exits in buildings of Group A-5 where all portions of the means of egress are essentially open to the outside need not be enclosed.
3. Stairways serving and contained within a single residential dwelling unit or sleeping unit in Group R-1, R-2 or R-3 occupancies are not required to be enclosed.
4. Stairways that are not a required means of egress element are not required to be enclosed where such stairways comply with Section 707.2.
5. Stairways in open parking structures that serve only the parking structure are not required to be enclosed.
6. Stairways in Group I-3 occupancies, as provided for in Section 408.3.6, are not required to be enclosed.
7. Means of egress stairways as required by Section 410.5.3 are not required to be enclosed.
8. In other than Group H and I occupancies, a maximum of 50 percent of egress stairways serving one adjacent floor are not required to be enclosed, provided at least two means of egress are provided from both floors served by the unenclosed stairways. Any two such interconnected floors shall not be open to other floors. Unenclosed exit stairways shall be remotely located as required in Section 1015.2.
9. In other than Group H and I occupancies, interior egress stairways serving only the first and second stories of a building equipped throughout with an automatic sprinkler system in accordance with Section 903.3.1.1 are not required to be enclosed, provided at least two means of egress are provided from both floors served by the unenclosed stairways. Such interconnected stories shall not be open to other stories. Unenclosed exit stairways shall be remotely located as required in Section 1015.2.

Q: Given a building on grade, without a basement, see Attachment No. 1 Multi-Story Building. Is the second story of this building the first story above the lowest level of exit discharge?

A: No. When determining stories above the lowest level of exit discharge, a level, or floor level, is not a story. A "level" is the horizontal plane that is part of a story, not the entire story height. A "story" is the vertical space between the upper surface of one floor level and the upper surface of the floor level next above or below.

The level of exit discharge as shown in attached drawing of a Multi-Story Building is at elevation 0.0 feet, the first floor level of the building is also at elevation 0.0 feet; therefore, the level of exit discharge and the first floor level of the building are at the same elevation. The first story of the building begins at elevation 0.0 feet (first floor) and extends to elevation 10.0 feet (second floor). The first story of the building is the first story above the level of exit discharge.

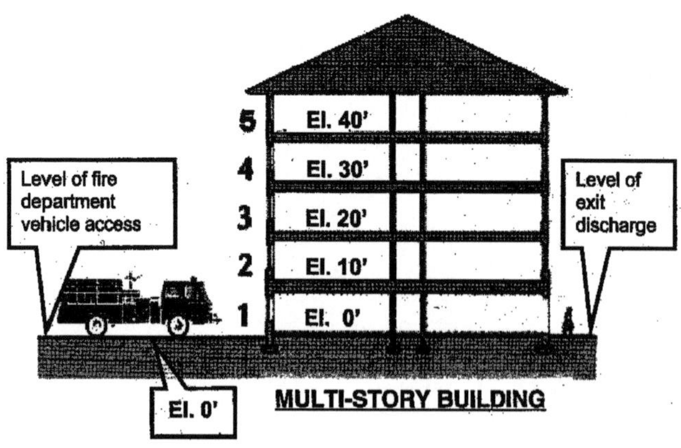

Attachment No. 1

Q: Are scissor stairs (e.g., two flights of stairs) allowed in the same enclosure under Section 1020.1?

A: Yes. Section 1020.1 does not limit the number of stairways allowed within a single enclosure. However, if more than one stairway is located in a single enclosure without being completely separated by a fire separation assembly with a fire-resistance rating as required for exits, the stairways would only be considered one exit for the purpose of the required number of exits.

Q: An automotive service shop has below-grade service areas where employees perform oil changes and other minor maintenance services. The below-grade areas are open to the grade-level service bays, thus providing access to the underside of the vehicles without requiring them to be lifted into the air. A stairway is provided to gain access to and egress from this below-grade service area. Is the stairway from this below-grade service level required to be enclosed in fire-resistance-rated construction in accordance with Section 1020.1?

A: No. Section 1020.1 requires exit stairways that connect two or more stories to be enclosed in fire-resistance-rated construction. The enclosure requirements would only apply to this below-grade service level if it is, in fact, a story. Due to the openness between the adjacent service levels, the below-grade area would be more appropriately regulated similar to a mezzanine rather than a story. A mezzanine is considered part of the same story that it serves, and egress stairways from the mezzanine level are not required to be enclosed in fire-resistance-rated construction in accordance with Exception 9 to Section 707.2.

Therefore, if the below-grade service level is in compliance with the applicable provisions of Section 505, the egress stairway serving the mezzanine level is not required to be enclosed in fire-resistance-rated construction.

1020.1.1 Openings and penetrations. Exit enclosure opening protectives shall be in accordance with the requirements of Section 715.

Except as permitted in Section 402.4.6, openings in exit enclosures other than unprotected exterior openings shall be limited to those necessary for exit access to the enclosure from normally occupied spaces and for egress from the enclosure.

Where interior exit enclosures are extended to the exterior of a building by an exit passageway, the door assembly from the exit enclosure to the exit passageway shall be protected by a fire door assembly conforming to the requirements in Section 715.4. Fire door assemblies in exit enclosures shall comply with Section 715.4.4.

Elevators shall not open into an exit enclosure.

Q: Are all doors, except corridor doors, prohibited by Section 1020.1.1 from opening into an exit stairway?

A: No. The intent of Section 1020.1.1 is to permit only doors that are part of an element of a means of egress to open into stairway enclosures. Rooms such as janitor's closets, mechanical equipment rooms, closets and other normally unoccupied spaces are prohibited from having doors opening into the stairway enclosure. Thus, rooms that would usually be occupied can have doors opening directly into the stairway enclosure. For buildings in which two or more exits are required, the means of egress from such occupied rooms must provide access to the other required exits in a manner that does not require passing through the exit stairway in question.

1020.1.2 Penetrations. Penetrations into and openings through an exit enclosure are prohibited except for required exit doors, equipment and ductwork necessary for independent pressurization, sprinkler piping, standpipes, electrical raceway for fire department communication and electrical raceway serving the exit enclosure and terminating at a steel box not exceeding 16 square inches ($0.010 m^2$). Such penetrations shall be protected in accordance with Section 712. There shall be no penetrations or communication openings, whether protected or not, between adjacent exit enclosures.

Q: Is a hot water unit heater permitted to be installed within an exit enclosure?

A: Yes. Portions of a buildings service systems that serve only the exit enclosure are permitted to penetrate the exit enclosure that is being served. The hot water unit heater cannot obstruct the path of egress travel, or reduce the clear width of the stair or landing. The boiler that provides the hot water for the unit heater must be located outside the exit enclosure.

Q: Is it the intent of Section 1020.1.2 to prohibit hose connections on standpipes from being located in the stairway enclosure?

A: No. The code provision requiring an exit stairway enclosure to not be used for any purpose other than means of egress is intended to prohibit hazards to means of egress and also to restrict the possibility of locating a fuel load within the enclosure. Use of the enclosure as a "staging area" by a fire department does not represent "use" of a stairway enclosure intended to be restricted by this section. Furthermore, NFPA 14, the standard referenced by the code, specifically requires, with certain exceptions, that standpipe risers be located within fire-resistance-rated stairway enclosures.

1020.1.3 Ventilation. Equipment and ductwork for exit enclosure ventilation as permitted by Section 1020.1.2 shall comply with one of the following items:

1. Such equipment and ductwork shall be located exterior to the building and shall be directly connected to the exit en-

closure by ductwork enclosed in construction as required for shafts.

2. Where such equipment and ductwork is located within the exit enclosure, the intake air shall be taken directly from the outdoors and the exhaust air shall be discharged directly to the outdoors, or such air shall be conveyed through ducts enclosed in construction as required for shafts.
3. Where located within the building, such equipment and ductwork shall be separated from the remainder of the building, including other mechanical equipment, with construction as required for shafts.

In each case, openings into the fire-resistance-rated construction shall be limited to those needed for maintenance and operation and shall be protected by opening protectives in accordance with Section 715 for shaft enclosures.

Exit enclosure ventilation systems shall be independent of other building ventilation systems.

Q: Are ducts permitted to penetrate the fire-resistance-rated enclosure walls of an exit stairway for purposes of heating, ventilating and cooling the stair enclosure?

A: Yes. Section 716.5.3 allows ducts to service the stairway enclosure when permitted by Section 1020.1.2. (Note: Section 1020.1.1 restricts penetrations to only those serving the exit itself).

SECTION 1023
EXTERIOR EXIT RAMPS AND STAIRWAYS

1023.6 Exterior ramps and stairway protection. Exterior exit ramps and stairways shall be separated from the interior of the building as required in Section 1020.1. Openings shall be limited to those necessary for egress from normally occupied spaces.

Exceptions:

1. Separation from the interior of the building is not required for occupancies, other than those in Group R-1 or R-2, in buildings that are no more than two stories above grade plane where the level of exit discharge is the first story above grade plane.
2. Separation from the interior of the building is not required where the exterior ramp or stairway is served by an exterior ramp and/or balcony that connects two remote exterior stairways or other approved exits with a perimeter that is not less than 50 percent open. To be considered open, the opening shall be a minimum of 50 percent of the height of the enclosing wall, with the top of the openings no less than 7 feet (2134 mm) above the top of the balcony.
3. Separation from the interior of the building is not required for an exterior ramp or stairway located in a building or structure that is permitted to have unenclosed interior stairways in accordance with Section 1020.1.
4. Separation from the interior of the building is not required for exterior ramps or stairways connected to open-ended corridors, provided that Items 4.1 through 4.4 are met:
    - 4.1. The building, including corridors and ramps and/or stairs, shall be equipped throughout with an automatic sprinkler system in accordance with Section 903.3.1.1 or 903.3.1.2.
    - 4.2. The open-ended corridors comply with Section 1017.
    - 4.3. The open-ended corridors are connected on each end to an exterior exit ramp or stairway complying with Section 1023.
    - 4.4. At any location in an open-ended corridor where a change of direction exceeding 45 degrees (0.79 rad) occurs, a clear opening of not less than 35 square feet (3.3 m^2) or an exterior ramp or stairway shall be provided. Where clear openings are provided, they shall be located so as to minimize the accumulation of smoke or toxic gases.

Q: Are exterior exit stairs required to be separated with fire-resistive-rated construction from open-ended corridors?

A: No. Exception 4 of Section 1023.6 defines specific provisions that permit the use of exterior stairs connected to open-ended corridors. An exterior stairway need not be separated from the interior of the building or from the open-ended corridor where both comply with the provisions in Items 4.1 through 4.4.

SECTION 1024
EXIT DISCHARGE

1024.1 General. Exits shall discharge directly to the exterior of the building. The exit discharge shall be at grade or shall provide direct access to grade. The exit discharge shall not reenter a building.

Exceptions:

1. A maximum of 50 percent of the number and capacity of the exit enclosures is permitted to egress through areas on the level of discharge provided all of the following are met:
    - 1.1. Such exit enclosures egress to a free and unobstructed way to the exterior of the building,

which way is readily visible and identifiable from the point of termination of the exit enclosure.

1.2. The entire area of the level of discharge is separated from areas below by construction conforming to the fire-resistance rating for the exit enclosure.

1.3. The egress path from the exit enclosure on the level of discharge is protected throughout by an approved automatic sprinkler system. All portions of the level of discharge with access to the egress path shall either be protected throughout with an automatic sprinkler system installed in accordance with Section 903.3.1.1 or 903.3.1.2, or separated from the egress path in accordance with the requirements for the enclosure of exits.

2. A maximum of 50 percent of the number and capacity of the exit enclosures is permitted to egress through a vestibule provided all of the following are met:

   2.1. The entire area of the vestibule is separated from areas below by construction conforming to the fire-resistance rating for the exit enclosure.

   2.2. The depth from the exterior of the building is not greater than 10 feet (3048 mm) and the length is not greater than 30 feet (9144 mm).

   2.3. The area is separated from the remainder of the level of exit discharge by construction providing protection at least the equivalent of approved wired glass in steel frames.

   2.4. The area is used only for means of egress and exits directly to the outside.

3. Stairways in open parking garages complying with Section 1020.1, Exception 5, are permitted to egress through the open parking garage at the level of exit discharge.

Q: Is the "required discharge capacity" of an exit discharge, as stated in Section 1024.2, to be determined using Section 1004.1?

A: Yes. The exit discharge is one component of a means of egress and must be sized to carry adequately the occupant load served. Table 1004.1.1 would be applied to determine that the width required to provide adequate capacity is achieved. Corridor and stairway width in accordance with Section 1005.1 would also be used to determine that the minimum width of the exit discharge is achieved.

Q: An enclosed mezzanine with more than ten occupants is provided with one enclosed exit stairway and one unenclosed stairway in accordance with Exception 2 of Section 505.4. Does Section 1024.1 prohibit both means of egress stairways from discharging into the same space at the floor below where the space qualifies as a lobby in accordance with Section 1024.1, Exception 1?

A: No. The provisions of Section 1024.1 are applicable to multiple exit elements that discharge into the same space within the building. An unenclosed stairway from the mezzanine is not considered an exit, but rather an element of exit access. The unenclosed stairway is not regulated by Section 1024.1. Therefore, both means of egress stairways that serve the mezzanine are permitted to discharge into the common lobby.

Q: Does Section 1024.1 require that a public way meet the means of egress requirements in the code relative to slip resistance, levelness, curbs, steps, illumination, etc.?

A: No. The means of egress exit elements of a building or structure must discharge directly to a public way or to a space serving as the exit discharge which ultimately leads to a public way. While the code provides specific requirements for the exit discharge leading to the public way, the code does not identify specific requirements for a public way. The definition of "Public way" establishes that it is limited to public use and has a clear width and a clear height of at least 10 feet. All other attributes of a public way are outside the scope of the code.

The intent of the means of egress requirements leading to a public way is to provide the occupants with a safe area where they are free from the immediate effects of a fire or other emergency situation within the building or structure. The code does not contain requirements for slip resistance, levelness, curbs, steps, illumination, etc., for public streets, public alleys or open parcels of land legally limited to public use.

SECTION 1026
EMERGENCY ESCAPE AND RESCUE

1026.1 General. In addition to the means of egress required by this chapter, provisions shall be made for emergency escape and rescue in Group R and I-1 occupancies. Basements and sleeping rooms below the fourth story above grade plane shall have at least one exterior emergency escape and rescue opening in accordance with this section. Where basements contain one or more sleeping rooms, emergency egress and rescue openings shall be required in each sleeping room, but shall not be required in adjoining areas of the basement. Such openings shall open directly into a public way or to a yard or court that opens to a public way.

Exceptions:

1. In other than Group R-3 occupancies, buildings equipped throughout with an approved automatic sprinkler system in accordance with Section 903.3.1.1 or 903.3.1.2.

2. In other than Group R-3 occupancies, sleeping rooms provided with a door to a fire-resistance-rated corridor having access to two remote exits in opposite directions.
3. The emergency escape and rescue opening is permitted to open onto a balcony within an atrium in accordance with the requirements of Section 404, provided the balcony provides access to an exit and the dwelling unit or sleeping unit has a means of egress that is not open to the atrium.
4. Basements with a ceiling height of less than 80 inches (2032 mm) shall not be required to have emergency escape and rescue windows.
5. High-rise buildings in accordance with Section 403.
6. Emergency escape and rescue openings are not required from basements or sleeping rooms that have an exit door or exit access door that opens directly into a public way or to a yard, court or exterior exit balcony that opens to a public way.
7. Basements without habitable spaces and having no more than 200 square feet (18.6 m^2) in floor area shall not be required to have emergency escape windows.

Q: A multistory building with an occupancy in Group R or I-1 has an exit access corridor and exit stairways separated from an atrium. If the sleeping rooms below the fourth story have windows of the minimum required dimensions and area opening onto the atrium, which then provides direct access to the outside, is the intent of Exception 3 to Section 1026.1 satisfied?

A: Yes. The purpose of the emergency escape window is to provide an alternative means of escape or rescue in the event that the required means of egress is not available. If the required means of egress from the sleeping room does not directly communicate with the atrium space, the likelihood of both routes blocked by the same incident is remote. On this basis, the intent of Exception 3 to Section 1026.1 has been satisfied.

Q: In a building with an occupancy in Group R or I-1, which is required to have a minimum of two exits per floor, would windows complying with the requirements of Section 1026.2 be considered acceptable as one of the required exits?

A: No. An emergency escape window is required in addition to the minimum number of required exits. The window is intended to serve the special function of emergency escape or rescue for sleeping rooms only and is not intended to serve as a substitution for the required exit facilities for a floor area.

1026.2 Minimum size. Emergency escape and rescue openings shall have a minimum net clear opening of 5.7 square feet (0.53 m^2).

Exception: The minimum net clear opening for emergency escape and rescue grade-floor openings shall be 5 square feet (0.46 m^2).

Q: What constitutes "grade-floor" openings as specified in the Exception to Section 1026.2?

A: The intent of the exception is to permit a reduction in the minimum net clear opening from 5.7 square feet to 5 square feet for those windows that can be accessed from the adjoining grade without the use of a ladder.

CHAPTER 11

ACCESSIBILITY

SECTION 1103
SCOPING REQUIREMENTS

1103.1 Where required. Sites, buildings, structures, facilities, elements and spaces, temporary or permanent, shall be accessible to persons with physical disabilities.

1103.2 General exceptions. Sites, buildings, structures, facilities, elements and spaces shall be exempt from this chapter to the extent specified in this section.

Q: Are sinks that are provided in areas other than toilet rooms and kitchens required to be accessible?

A: Yes. Section 1103 states that all structures are required to be fully accessible unless specifically exempted. While Section 1109.2 provides specific requirements for toilet and bathing facilities, it does not address individual sinks in other building locations. The intent of Chapter 11 is to require buildings and structures, including all of their associated sites and facilities, to be fully accessible to persons with physical disabilities unless specifically exempted.

Q: In structures containing occupancies of Groups A, I and R, which are provided with the required accessible elements in accordance with Section 1108 on an accessible level, is an accessible route required to other levels?

A: No. In new construction, an accessible route is not required to every level, provided specific exceptions are met. Section 1103.1 states that all structures are required to be fully accessible. If a level contains only elements specifically identified in Section 1108 (e.g., assembly seating, dwelling units and sleeping units) and the accessibility requirements are met by the elements provided on an accessible level, then an accessible route to other levels not containing any accessible elements is not required. This is specifically stated in Section 1104.3, which requires the accessible route to connect accessible elements. For small areas, Section 1104.4, Exception 1, would be applicable. If accessible elements were required on levels above and below the accessible level, but the aggregate area was less than 3,000 square feet, and the level did not contain the functions described in the exception (i.e., offices of health care providers, passenger transportation facilities, airports and multitenant facilities of Group M), an accessible route would not be required to those areas.

SECTION 1104
ACCESSIBLE ROUTE

1104.4 Multilevel buildings and facilities. At least one accessible route shall connect each accessible level, including mezzanines, in multilevel buildings and facilities.

Exceptions:

1. An accessible route is not required to stories and mezzanines above and below accessible levels that have an aggregate area of not more than 3,000 square feet (278.7 m^2). This exception shall not apply to:
    - 1.1. Multiple tenant facilities of Group M occupancies containing five or more tenant spaces;
    - 1.2. Levels containing offices of health care providers (Group B or I); or
    - 1.3. Passenger transportation facilities and airports (Group A-3 or B).
2. In Group A, I, R and S occupancies, levels that do not contain accessible elements or other spaces required by Section 1107 or 1108 are not required to be served by an accessible route from an accessible level.
3. In air traffic control towers, an accessible route is not required to serve the cab and the floor immediately below the cab.
4. Where a two-story building or facility has one story with an occupant load of five or fewer persons that does not contain public use space, that story shall not be required to be connected by an accessible route to the story above or below.

Q: Given a two-story free-standing building, Group B occupancy, with a building area of 8,800 square feet per floor. The building is divided into four two-story independent tenant spaces. There are no shared facilities, such as mechanical and plumbing facilities, toilet rooms, janitor's closets, or other common use areas connecting the four tenant spaces. The four tenant spaces are functionally self-sufficient.

Both floors in all of the tenant spaces are required to be accessible. The main entrance level is fully accessible, including an accessible route. An interior exit access stairway within each tenant space connects the upper and lower levels. The nonentrance level for each tenant is 3,000 square feet or less.

Does Exception 1 of Section 1104.4 permit the 3,000 square foot area limit for an accessible route to be applied independently to each of the four unconnected tenant spaces?

A: Yes. Individual tenant spaces that do not share communicating or common use areas with adjacent tenant spaces may be considered separate "facilities." Therefore, each individual tenant space is permitted to be evaluated individually with regards to the accessible route requirements in Section 1104.4, Exception 1. Levels with an aggregate area not more than 3,000 square feet within a single tenant space are not required to be connected with an accessible route.

Q: Under the provisions of Exception 2 to Section 1104.4 of the *International Building Code*, in a Group S occupancy, are levels that do not contain any accessible elements required to be served by an accessible route from a required accessible level?

A: No. The provisions in Exception 2 to Section 1104.4 only apply to occupancy Groups A, I, R and S limited to the specific items addressed in Section 1107 and Section 1108.

In a Group S storage facility that provides self-service storage, Section 1108.3 requires a minimum number of the individual self-service storage spaces to be accessible. Where individual self-service storage spaces are located on a level(s) above and/or below a required accessible level that level(s) must be connected by an accessible route. However, Section 1108.3.1 does not require accessible individual self-service storage spaces to be dispersed throughout the building. Where a level(s) is not provided with accessible individual self-service storage spaces an accessible route is not required to connect that level(s) with a required accessible level.

SECTION 1106
PARKING AND PASSENGER LOADING FACILITIES

1106.2 Groups R-2 and R-3. Two percent, but not less than one, of each type of parking space provided for occupancies in Groups R-2 and R-3, which are required to have Accessible, Type A or Type B dwelling or sleeping units, shall be accessible. Where parking is provided within or beneath a building, accessible parking spaces shall also be provided within or beneath the building.

Q: Are private garages that are attached to and serve specific individual Group R-2 dwelling units required to comply with the 2 percent minimum for each type of parking space?

A: Yes. The *International Building Code* does not require buildings and facilities to be provided with parking spaces; however, where parking spaces are provided for a Group R-2 occupancy that is required to have Type A or Type B dwelling units, at least 2 percent of each type of parking provided must be accessible. This includes individual private garages, where individual private garages are one of the types of parking provided.

CHAPTER 12

INTERIOR ENVIRONMENT

SECTION 1208
INTERIOR SPACE DIMENSIONS

1208.2 Minimum ceiling heights. Occupiable spaces, habitable spaces and corridors shall have a ceiling height of not less than 7 feet 6 inches (2286 mm). Bathrooms, toilet rooms, kitchens, storage rooms and laundry rooms shall be permitted to have a ceiling height of not less than 7 feet (2134 mm).

Exceptions:

1. In one- and two-family dwellings, beams or girders spaced not less than 4 feet (1219 mm) on center and projecting not more than 6 inches (152 mm) below the required ceiling height.
2. If any room in a building has a sloped ceiling, the prescribed ceiling height for the room is required in one-half the area thereof. Any portion of the room measuring less than 5 feet (1524 mm) from the finished floor to the ceiling shall not be included in any computation of the minimum area thereof.
3. Mezzanines constructed in accordance with Section 505.1.

Q: Is a mezzanine that is added to an existing room and used only for storage required to comply with the minimum ceiling height requirements of Section 1208.2 regarding occupiable rooms?

A: No. An "occupiable space" is defined in Chapter 2 as room or enclosed space designed for human occupancy or for occupants engaged in labor. If the mezzanine is used strictly for storage, and not as an area where occupants are engaged in labor, it does not fall within the definition of an "Occupiable space," and therefore is not subject to minimum ceiling height requirements. However, if the area below the mezzanine is a habitable or occupiable space, the minimum required ceiling height must be maintained in that area.

CHAPTER 14

EXTERIOR WALL COVERINGS

SECTION 1403 PERFORMANCE REQUIREMENTS

1403.2 Weather protection. Exterior walls shall provide the building with a weather-resistant exterior wall envelope. The exterior wall envelope shall include flashing, as described in Section 1405.3. The exterior wall envelope shall be designed and constructed in such a manner as to prevent the accumulation of water within the wall assembly by providing a water-resistive barrier behind the exterior veneer, as described in Section 1404.2 and a means for draining water that enters the assembly to the exterior of the veneer. Protection against condensation in the exterior wall assembly shall be provided in accordance with the *International Energy Conservation Code.*

Exceptions:

1. A weather-resistant exterior wall envelope shall not be required over concrete or masonry walls designed in accordance with Chapters 19 and 21, respectively.
2. Compliance with the requirements for a means of drainage, and the requirements of Sections 1405.2 and 1405.3, shall not be required for an exterior wall envelope that has been demonstrated through testing to resist wind-driven rain, including joints, penetrations and intersections with dissimilar materials, in accordance with ASTM E 331 under the following conditions:
    - 2.1. Exterior wall envelope test assemblies shall include at least one opening, one control joint, one wall/eave interface and one wall sill. All tested openings and penetrations shall be representative of the intended end-use configuration.
    - 2.2. Exterior wall envelope test assemblies shall be at least 4 feet by 8 feet (1219 mm by 2438 mm) in size.
    - 2.3. Exterior wall envelope assemblies shall be tested at a minimum differential pressure of 6.24 pounds per square foot (psf) (0.297 kN/m^2).
    - 2.4. Exterior wall envelope assemblies shall be subjected to a minimum test exposure duration of 2 hours.

The exterior wall envelope design shall be considered to resist wind-driven rain where the results of testing indicate that water did not penetrate control joints in the exterior wall envelope, joints at the perimeter of openings or intersections of terminations with dissimilar materials.

Q: Does Section 1403.2 require attics and crawl spaces to be enclosed with exterior walls protected with weather-resistant coverings?

A: Yes. The weather protection requirements contained in the code are intended to provide all exterior walls with approved weather-resistant coverings. These requirements are equally applicable to all enclosed building spaces, regardless of whether the walls enclose habitable/occupiable areas or other unoccupied spaces.

SECTION 1405 INSTALLATION OF WALL COVERINGS

1405.3 Flashing. Flashing shall be installed in such a manner so as to prevent moisture from entering the wall or to redirect it to the exterior. Flashing shall be installed at the perimeters of exterior door and windows assemblies, penetrations and terminations of exterior wall assemblies, exterior wall intersections with roofs, chimneys, porches, decks, balconies and similiar projections and at built-in gutters and similiar locations where moisture could enter the wall. Flashing with projecting flanges shall be installed on both sides and the ends of copings, under sills and continuously above projecting trim.

Q: Do typical metal gutters and leaders require the same flashing protection as "built-in gutters?"

A: No. Sections 1405.3 and 1503.2 provide flashing requirements for intersections where exterior water is likely to enter the building. The flashing requirements for built-in gutters only refer to gutters integrated into the building structure, thus creating a vertical-to-horizontal intersection which is subject to water entry. Generally, attached metal gutters receive rainwater from roofs protected with special eave underlayment and metal drip edges. Therefore, additional flashing is not required.

SECTION 1406 COMBUSTIBLE MATERIALS ON THE EXTERIOR SIDE OF EXTERIOR WALLS

1406.2.2 Architectural trim. In buildings of Types I, II, III, and IV construction that do not exceed three stories or 40 feet (12 192 mm) in height above grade plane, exterior wall coverings shall be permitted to be constructed of wood where permitted by Section 1405.4 or other equivalent combustible material. Combustible exterior wall coverings, other than fire-retardant-treated wood complying with Section 2303.2 for exterior installation, shall not exceed 10 percent of an exterior wall surface area where the fire separation distance is 5 feet

(1524 mm) or less. Architectural trim that exceeds 40 feet (12 192 mm) in height above grade plane shall be constructed of approved noncombustible materials and shall be secured to the wall with metal or other approved noncombustible brackets.

Q: Is combustible exterior wall finish permitted to be installed on exterior walls that are required to be constructed of noncombustible materials?

A: Yes. Section 1406.2 regulates the use of combustible exterior wall finish. Sections 1406.2.2 and 1406.2.3 regulate the use of such finish and trim and are equally applicable to both combustible and noncombustible walls.

CHAPTER 15

ROOFS AND ROOF STRUCTURES

SECTION 1509 ROOFTOP STRUCTURES

1509.2 Penthouses. A penthouse or other projection above the roof in structures of other than Type I construction shall not exceed 28 feet (8534 mm) above the roof where used as an enclosure for tanks or for elevators that run to the roof and in all other cases shall not extend more than 18 feet (3658 mm) above the roof. The aggregate area of penthouses and other rooftop structures shall not exceed one-third the area of the supporting roof. A penthouse, bulkhead, or any other similar projection above the roof shall not be used for purposes other than shelter of mechanical equipment or shelter of vertical shaft openings in the roof. Provisions, such as louvers, louver blades or flashing, shall be made to protect the mechanical equipment and the building interior from the elements. Penthouses or bulkheads used for purposes other than permitted by this section shall conform to the requirements of this code for an additional story. The restrictions of this section shall not prohibit the placing of wood flagpoles or similar structures on the roof of any building.

Q: Does a penthouse contribute to the number of stories above grade or to the height in feet of a building?

A: No. Section 1502 defines a "Penthouse" as an enclosed structure located above the roof of the building, occupying not more than $33^1/_3$ percent of the roof area. Section 1509.2 states that a penthouse complying with these provisions would be considered a part of the next lower story. Therefore, a penthouse does not contribute to either the number of stories above grade or to the height in feet of a building.

CHAPTER 16

STRUCTURAL DESIGN

SECTION 1605 LOAD COMBINATIONS

1605.3.1 Basic load combinations. Where allowable stress design (working stress design), as permitted by this code, is used, structures and portions thereof shall resist the most critical effects resulting from the following combinations of loads:

$D + F$ **(Equation 16-8)**

$D + H + F + L + T$ **(Equation 16-9)**

$D + H + F + (L_r \text{ or } S \text{ or } R)$ **(Equation 16-10)**

$D + H + F + 0.75(L + T) + 0.75(L_r \text{ or } S \text{ or } R)$ **(Equation 16-11)**

$D + H + F + (W \text{ or } 0.7E)$ **(Equation 16-12)**

$D + H + F + 0.75(W \text{ or } 0.7E) + 0.75L + 0.75(L_r \text{ or } S \text{ or } R)$ **(Equation 16-13)**

$0.6D + W + H$ **(Equation 16-14)**

$0.6D + 0.7E + H$ **(Equation 16-15)**

Exceptions:

1. Crane hook loads need not be combined with roof live load or with more than three-fourths of the snow load or one-half of the wind load.
2. Flat roof snow loads of 30 psf (1.44 kN/m^2) or less need not be combined with seismic loads. Where flat roof snow loads exceed 30 psf (1.44 kN/m^2), 20 percent shall be combined with seismic loads.

Q: Does Exception 2 to Section 1605.3.1 apply to Equation 16-15 even though this load combination does not include the snow term?

A: No. This is an exception to the requirement to combine the effects of snow loads and earthquake loads as would otherwise be required by Equation 16-13.

1605.3.2 Alternative basic load combinations. In lieu of the basic load combinations specified in Section 1605.3.1, structures and portions thereof shall be permitted to be designed for the most critical effects resulting from the following combinations. When using these alternative basic load combinations that include wind or seismic loads, allowable stresses are permitted to be increased or load combinations reduced where permitted by the material chapter of this code or the referenced standards. For load combinations that include the counteracting effects of dead and wind loads, only two-thirds of the minimum dead load likely to be in place during a design wind event shall be used. Where wind loads are calculated in accordance with Chapter 6 of ASCE 7, the coefficient ω in the following equations shall be taken as 1.3. For other wind loads, ω shall be taken as 1. When using these alternative load combinations to evaluate sliding, overturning and soil bearing at the soil-structure interface, the reduction of foundation overturning from Section 12.13.4 in ASCE 7 shall not be used. When using these alternative basic load combinations for proportioning foundations for loadings, which include seismic loads, the vertical seismic load effect, E_v, in Equation 12.4-4 of ASCE 7 is permitted to be taken equal to zero.

$D + L + (L_r \text{ or } S \text{ or } R)$ **(Equation 16-16)**

$D + L + (\omega W)$ **(Equation 16-17)**

$D + L + \omega W + S/2$ **(Equation 16-18)**

$D + L + S + \omega W/2$ **(Equation 16-19)**

$D + L + S + E/1.4$ **(Equation 16-20)**

$0.9D + E/1.4$ **(Equation 16-21)**

Exceptions:

1. Crane hook loads need not be combined with roof live loads or with more than three-fourths of the snow load or one-half of the wind load.
2. Flat roof snow loads of 30 psf (1.44 kN/m^2) or less need not be combined with seismic loads. Where flat roof snow loads exceed 30 psf (1.44 kN/m^2), 20 percent shall be combined with seismic loads.

Q: Does Exception 2 to Section 1605.3.2 apply to Equation 16-21 even though this load combination does not include the snow term?

A: No. This is an exception to the requirement to combine the effects of snow loads and earthquake loads as would otherwise be required by Equation 16-20.

SECTION 1607 LIVE LOADS

1607.3 Uniform live loads. The live loads used in the design of buildings and other structures shall be the maximum loads expected by the intended use or occupancy but shall in no case be less than the minimum uniformly distributed unit loads required by Table 1607.1.

TABLE 1607.1, see page 78.

Q: Can the required design live loads of Table 1607.1 for balconies be reduced for extremely small or narrow ornamental balconies or narrow balconies that are not available for normal use?

A: Yes. The required design live loads are intended to apply to balconies that accommodate day-to-day usage demands when available to the building occupants for uses generally associated with balconies. If, in the opinion of the code official, the size, configuration or availability of the balcony does not lend itself to usage as the term "balcony" is usually applied, reduced or lesser values of design live loads may be utilized, representing the actual anticipated loads.

Q: Does Table 1607.1 require subflooring to be installed on attic joists and the bottom chord of roof trusses to accommodate the minimum 20 pound per square foot uniform live load?

A: No. The intent of Table 1607.1, Item 28, is to require that attic joists or the bottom chords of trusses be designed to support a specified live load so as to produce a safe and stable structural element. Although these structural members are required to be capable of supporting a minimum live load, the code does not mandate that subflooring be installed nor does it prescribe a specific method of supporting the loads.

1607.9 Reduction in live loads. Except for roof uniform live loads, all other minimum uniformly distributed live loads, L_o, in Table 1607.1 are permitted to be reduced in accordance with Section 1607.9.1 or 1607.9.2.

Q: Can the permissible live load reduction of Section 1607.9 be applied to buildings or structures or portions thereof where the actual live loads are known or a high probability exists that the full live load will occur?

A: No. Section 1607.9 provisions are based on the low probability of structural members of buildings being fully loaded to the minimum uniformly distributed live loads as shown in Table 1607.1. Furthermore, Section 1607.3 directs the designer to utilize the greater live loads produced by the intended occupancy but not less than the minimum uniformly distributed live loads listed in Table 1607.1. Thus, where actual live loads will or are likely to occur based on the intended occupancy, live load reductions are not applicable to those members supporting such loads.

Q: Does the reference to tributary area for structural members include all floor areas that contribute to loading on the member?

A: Yes. Section 1607.9 provisions are based on the probabilities of simultaneous full live loading that occurs on all tributary areas. Section 1607.9 allows a reduced design live load for all members with an influence area ($K_{LL} A_T$) of 400 square feet or more, subject to the maximum reductions of Section 1607.9.1.1. Unless unusual circumstances exist that make full simultaneous live loading of all tributary areas probable, the total influence area may be considered as the accumulative tributary area of all floor areas supported by the structural member.

1607.9.1.3 Special occupancies. Live loads of 100 psf (4.79 kN/m^2) or less shall not be reduced in public assembly occupancies.

Q: Can the live load be reduced in public assembly occupancies where the live load is 100 psf or less?

A: No. Since it is likely that the load will be imposed over a large portion of the floor area, the live load of an assembly occupancy can not be reduced. Should the live load exceed 100 psf, Section 1607.9.1.1 would permit a reduction by a maximum of 20 percent of the live loads for members supporting two or more floors, provided the live load is not less than L as calculated in Section 1607.9.1.

1607.13 Interior walls and partitions. Interior walls and partitions that exceed 6 feet (1829 mm) in height, including their finish materials, shall have adequate strength to resist the loads to which they are subjected but not less than a horizontal load of 5 pounds per square foot (0.240 kN/m^2).

Q: Is it the intent of Section 1607.13 to restrict the spacing of wall studs so the finish provides a nonflexible wall surface?

A: No. The horizontal load of 5 psf is to be resisted by both the finish and the wall system supporting it. The maximum stud spacing in nonload-bearing walls is regulated by Section 2308.9.2.3 as 28 inches. Section 1607.13 is not intended to control the deflection of finishes but rather to control the required minimum finish material and wall strength.

SECTION 1609
WIND LOADS

1609.1 Applications. Buildings, structures and parts thereof shall be designed to withstand the minimum wind loads prescribed herein. Decreases in wind loads shall not be made for the effect of shielding by other structures.

Q: Are roof trusses, which are placed on top of a flat concrete roof slab to create a sloped roof, required to be capable of resisting the appropriate wind uplift loads for components and cladding specified in Chapter 16?

A: Yes. Section 1609.1 requires buildings, structures and parts thereof to be designed to withstand the appropriate wind loads described in Chapter 16.

SECTION 1611
RAIN LOADS

1611.3 Controlled drainage. Roofs equipped with hardware to control the rate of drainage shall be equipped with a secondary drainage system at a higher elevation that limits accumulation of water on the roof above that elevation. Such roofs shall be designed to sustain the load of rainwater that will accumulate on them to the elevation of the secondary drainage system plus the uniform load caused by water that rises above the inlet of the secondary drainage system at its design flow determined from Section 1611.1. Such roofs shall also be checked for ponding instability in accordance with Section 1611.2.

Q: Does Section 1611.3 require the roof to be designed for ponding instability and controlled drainage resulting from blockage of the primary roof drainage system where such blockage occurs at locations in the system other than at the primary surface drains?

A: Yes. The blockage of the primary roof drainage system could be caused by debris or ice at the entrance to the primary roof surface drains or could occur within the piping system itself as a result of a piping failure or blockage due to ice or other debris. Any secondary means of roof drainage must be independent of the primary means to avoid the possibility that both means of drainage are blocked simultaneously. Overflow scuppers, the roof edge and secondary roof drains are often utilized as a secondary means for controlled roof drainage.

TABLE 1607.1
MINIMUM UNIFORMLY DISTRIBUTED LIVE LOADS AND MINIMUM CONCENTRATED LIVE LOADS[g]

OCCUPANCY OR USE	UNIFORM (psf)	CONCENTRATED (lbs.)
1. Apartments (see residential)	—	—
2. Access floor systems Office use Computer use	 50 100	 2,000 2,000
3. Armories and drill rooms	150	—
4. Assembly areas and theaters Fixed seats (fastened to floor) Follow spot, projections and control rooms Lobbies Movable seats Stages and platforms	60 50 100 100 125	—
5. Balconies On one- and two-family residences only, and not exceeding 100 sq ft	100 60	—
6. Bowling alleys	75	—
7. Catwalks	40	300
8. Dance halls and ballrooms	100	—
9. Decks	Same as occupancy served[h]	—
10. Dining rooms and restaurants	100	—
11. Dwellings (see residential)	—	—
12. Cornices	60	—
13. Corridors, except as otherwise indicated	100	—
14. Elevator machine room grating (on area of 4 in^2)	—	300
15. Finish light floor plate construction (on area of 1 in^2)	— —	200
16. Fire escapes On single-family dwellings only	100 40	—
17. Garages (passenger vehicles only) Trucks and buses	40 Note a See Section 1607.6	
18. Grandstands (see stadium and arena bleachers)	—	—
19. Gymnasiums, main floors and balconies	100	—
20. Handrails, guards and grab bars	See Section 1607.7	
21. Hospitals Corridors above first floor Operating rooms, laboratories Patient rooms	 80 60 40	 1,000 1,000 1,000
22. Hotels (see residential)	—	—
23. Libraries Corridors above first floor Reading rooms Stack rooms	 80 60 150[b]	 1,000 1,000 1,000
24. Manufacturing Heavy Light	 250 125	 3,000 2,000
25. Marquees	75	—
26. Office buildings Corridors above first floor File and computer rooms shall be designed for heavier loads based on anticipated occupancy Lobbies and first-floor corridors Offices	 80 — 100 50	 2,000 — 2,000 2,000
27. Penal institutions Cell blocks Corridors	 40 100	—
28. Residential One- and two-family dwellings Uninhabitable attics without storage[i] Uninhabitable attics with limited storage[i, j, k] Habitable attics and sleeping areas All other areas except balconies and decks Hotels and multiple-family dwellings Private rooms and corridors serving them Public rooms and corridors serving them	 10 20 30 40 40 100	—
29. Reviewing stands, grandstands and bleachers	Note c	
30. Roofs All roof surfaces subject to maintenance workers Awnings and canopies Fabric construction supported by a lightweight rigid skeleton structure All other construction Ordinary flat, pitched, and curved roofs Primary roof members, exposed to a work floor Single panel point of lower chord of roof trusses or any point along primary structural members supporting roofs: Over manufacturing, storage warehouses, and repair garages All other occupancies Roofs used for other special purposes Roofs used for promenade purposes Roofs used for roof gardens or assembly purposes	 5 nonreduceable 20 20 Note 1 60 100	 300 2,000 300 Note 1

(continued)

TABLE 1607.1—continued
MINIMUM UNIFORMLY DISTRIBUTED LIVE LOADS AND MINIMUM CONCENTRATED LIVE LOADS[g]

OCCUPANCY OR USE	UNIFORM (psf)	CONCENTRATED (lbs.)
31. Schools		
Classrooms	40	1,000
Corridors above first floor	80	1,000
First-floor corridors	100	1,000
32. Scuttles, skylight ribs and accessible ceilings	—	200
33. Sidewalks, vehicular driveways and yards, subject to trucking	250[d]	8,000[e]
34. Skating rinks	100	—
35. Stadiums and arenas		—
Bleachers	100[c]	
Fixed seats (fastened to floor)	60[c]	
36. Stairs and exits		Note f
One- and two-family dwellings	40	
All other	100	
37. Storage warehouses (shall be designed for heavier loads if required for anticipated storage)		
Heavy	250	
Light	125	
38. Stores		
Retail		
First floor	100	1,000
Upper floors	75	1,000
Wholesale, all floors	125	1,000
39. Vehicle barriers	See Section 1607.7.3	
40. Walkways and elevated platforms (other than exitways)	60	—
41. Yards and terraces, pedestrians	100	—

For SI: 1 inch = 25.4 mm, 1 square inch = 645.16 mm^2,
1 square foot = 0.0929 m^2,
1 pound per square foot = 0.0479 kN/m^2, 1 pound = 0.004448 kN,
1 pound per cubic foot = 16 kg/m^3

a. Floors in garages or portions of buildings used for the storage of motor vehicles shall be designed for the uniformly distributed live loads of Table 1607.1 or the following concentrated loads: (1) for garages restricted to vehicles accommodating not more than nine passengers, 3,000 pounds acting on an area of 4.5 inches by 4.5 inches; (2) for mechanical parking structures without slab or deck which are used for storing passenger vehicles only, 2,250 pounds per wheel.

b. The loading applies to stack room floors that support nonmobile, double-faced library bookstacks, subject to the following limitations:
   1. The nominal bookstack unit height shall not exceed 90 inches;
   2. The nominal shelf depth shall not exceed 12 inches for each face; and
   3. Parallel rows of double-faced bookstacks shall be separated by aisles not less than 36 inches wide.

c. Design in accordance with the ICC *Standard on Bleachers, Folding and Telescopic Seating and Grandstands.*

d. Other uniform loads in accordance with an approved method which contains provisions for truck loadings shall also be considered where appropriate.

e. The concentrated wheel load shall be applied on an area of 20 square inches.

f. Minimum concentrated load on stair treads (on area of 4 square inches) is 300 pounds.

g. Where snow loads occur that are in excess of the design conditions, the structure shall be designed to support the loads due to the increased loads caused by drift buildup or a greater snow design determined by the building official (see Section 1608). For special-purpose roofs, see Section 1607.11.2.2.

h. See Section 1604.8.3 for decks attached to exterior walls.

i. Attics without storage are those where the maximum clear height between the joist and rafter is less than 42 inches, or where there are not two or more adjacent trusses with the same web configuration capable of containing a rectangle 42 inches high by 2 feet wide, or greater, located within the plane of the truss. For attics without storage, this live load need not be assumed to act concurrently with any other live load requirements.

j. For attics with limited storage and constructed with trusses, this live load need only be applied to those portions of the bottom chord where there are two or more adjacent trusses with the same web configuration capable of containing a rectangle 42 inches high by 2 feet wide or greater, located within the plane of the truss. The rectangle shall fit between the top of the bottom chord and the bottom of any other truss member, provided that each of the following criteria is met:
   i. The attic area is accessible by a pull-down stairway or framed opening in accordance with Section 1209.2, and
   ii. The truss shall have a bottom chord pitch less than 2:12.
   iii. Bottom chords of trusses shall be designed for the greater of actual imposed dead load or 10 psf, uniformly distributed over the entire span.

k. Attic spaces served by a fixed stair shall be designed to support the minimum live load specified for habitable attics and sleeping rooms.

l. Roofs used for other special purposes shall be designed for appropriate loads as approved by the building official.

CHAPTER 17

STRUCTURAL TEST AND SPECIAL INSPECTIONS

SECTION 1708
STRUCTURAL TESTING FOR SEISMIC RESISTANCE

1708.5 Seismic qualification of mechanical and electrical equipment. The registered design professional in responsible charge shall state the applicable seismic qualification requirements for designated seismic systems on the construction documents. Each manufacturer of designated seismic system components shall test or analyze the component and its mounting system or anchorage and submit a certificate of compliance for review and acceptance by the registered design professional in responsible charge of the design of the designated seismic system and for approval by the building official. Qualification shall be by an actual test on a shake table, by three-dimensional shock tests, by an analytical method using dynamic characteristics and forces, by the use of experience data (i.e., historical data demonstrating acceptable seismic performance) or by a more rigorous analysis providing for equivalent safety.

Q: Is the certificate of compliance verifying testing and analyzing of designated seismic system components required to be prepared by an approved agency?

A: Yes. The system/component manufacturer must submit the certificate of compliance, but the testing must be performed by an approved agency. The manufacturer must also apply the required label to the product as defined in Section 1702.1.

SECTION 1714
PRECONSTRUCTION LOAD TEST

1714.5.1 Exterior windows and doors. Exterior windows and sliding doors shall be tested and labeled as conforming to AAMA/WDMA/CSA101/I.S.2/A440. The label shall state the name of the manufacturer, the approved labeling agency and the product designation as specified in AAMA/WDMA/CSA101/I.S.2/A440. Exterior side-hinged doors shall be tested and labeled as conforming to AAMA/WDMA/CSA101/I.S.2/A440 or comply with Section 1714.5.2. Products tested and labeled as conforming to AAMA/WDMA/CSA 101/I.S.2/A440 shall not be subject to the requirements of Sections 2403.2 and 2403.3.

Q: Are exterior windows, sliding doors and side-hinged doors required to fully comply with all the mandatory requirements of AAMA/WDMA/CSA101/I.S.2/A440 that are applicable to exterior windows and glass doors manufactured using the referenced materials?

A: No. Aluminum, vinyl and wood exterior windows and glass doors are subject only to the provisions in the referenced standard AAMA/WDMA/CSA101/I.S.2/A440 as they apply to the design pressure rating. Section 1714.5 charges that exterior windows and doors shall comply with the design pressure rating as determined in one of the listed standards. There are no provisions requiring exterior windows and doors to comply with all the performance standards in the specifications.

The referenced standards are specifications that establish minimum performance standards for the design, materials, construction, operation and performance of exterior windows and doors.

CHAPTER 18

SOILS AND FOUNDATIONS

SECTION 1802
FOUNDATION AND SOILS INVESTIGATIONS

1802.1 General. Foundation and soils investigations shall be conducted in conformance with Sections 1802.2 through 1802.6. Where required by the building official, the classification and investigation of the soil shall be made by a registered design professional.

1802.2 Where required. The owner or applicant shall submit a foundation and soils investigation to the building official where required in Sections 1802.2.1 through 1802.2.7.

> **Exception:** The building official need not require a foundation or soils investigation where satisfactory data from adjacent areas is available that demonstrates an investigation is not necessary for any of the conditions in Sections 1802.2.1 through 1802.2.6.

1802.2.1 Questionable soil. Where the safe-sustaining power of the soil is in doubt, or where a load-bearing value superior to that specified in this code is claimed, the building official shall require that the necessary investigation be made. Such investigation shall comply with the provisions of Sections 1802.4 through 1802.6.

1802.2.2 Expansive soils. In areas likely to have expansive soil, the building official shall require soil tests to determine where such soils do exist.

1802.2.3 Ground-water table. A subsurface soil investigation shall be performed to determine whether the existing ground-water table is above or within 5 feet (1524 mm) below the elevation of the lowest floor level where such floor is located below the finished ground level adjacent to the foundation.

> **Exception:** A subsurface soil investigation shall not be required where waterproofing is provided in accordance with Section 1807.

1802.2.4 Pile and pier foundations. Pile and pier foundations shall be designed and installed on the basis of a foundation investigation and report as specified in Sections 1802.4 through 1802.6 and Section 1808.2.2.

1802.2.5 Rock strata. Where subsurface explorations at the project site indicate variations or doubtful characteristics in the structure of the rock upon which foundations are to be constructed, a sufficient number of borings shall be made to a depth of not less than 10 feet (3048 mm) below the level of the foundations to provide assurance of the soundness of the foundation bed and its load-bearing capacity.

1802.2.6 Seismic Design Category C. Where a structure is determined to be in Seismic Design Category C in accordance with Section 1613, an investigation shall be conducted, and shall include an evaluation of the following potential hazards resulting from earthquake motions: slope instability, liquefaction and surface rupture due to faulting or lateral spreading.

1802.2.7 Seismic Design Category D, E or F. Where the structure is determined to be in Seismic Design Category D, E or F, in accordance with Section 1613, the soils investigation requirements for Seismic Design Category C, given in Section 1802.2.6, shall be met, in addition to the following. The investigation shall include:

1. A determination of lateral pressures on basement and retaining walls due to earthquake motions.
2. An assessment of potential consequences of any liquefaction and soil strength loss, including estimation of differential settlement, lateral movement or reduction in foundation soil-bearing capacity, and shall address mitigation measures. Such measures shall be given consideration in the design of the structure and can include, but are not limited to, ground stabilization, selection of appropriate foundation type and depths, selection of appropriate structural systems to accommodate anticipated displacements or any combination of these measures. The potential for liquefaction and soil strength loss shall be evaluated for site peak ground acceleration magnitudes and source characteristics consistent with the design earthquake ground motions. Peak ground acceleration shall be determined from a site-specific study taking into account soil amplification effects, as specified in Chapter 21 of ASCE.

> **Exception:** A site-specific study need not be performed provided that peak ground acceleration equal to $S_{DS}/2.5$ is used, where S_{DS} is determined in accordance with Section 21.2.1 of ASCE 7.

Q: Subsequent to the approval of a permit, but during foundation system construction, unusually poor soil conditions are encountered that significantly differ from the initial soil analysis that accompanied the permit application. Can the code official require further foundation investigations?

A: Yes. Section 105.4 states that the permit is not to be construed as authority to violate, cancel or set aside any code provisions. Therefore, additional information as to the adequacy of the design may be required after the permit is issued.

Section 1802.2.1 states: "Where the safe sustaining power of the soil is in doubt, or where a load-bearing value superior to that specified in this code is claimed, the building official shall require that the necessary borings or tests be made." The tests should be conducted by a registered design professional experienced in foundation design. Based on the tests, the design professional should make a recommendation to the code official as to the adequacy of the proposed foundation design.

Q: In multiple-building developments, under Section 1802.2 is a soil investigation required for each structure?

A: No. The exception to Section 1802.2 states that the code official can approve the use of soil data developed based on soil information from immediately adjacent areas. If there is any doubt regarding the soil characteristics at a particular site, the code official shall require an additional soil investigation in accordance with Section 1802.2.1.

SECTION 1805 FOOTINGS AND FOUNDATIONS

1805.2.1 Frost protection. Except where otherwise protected from frost, foundation walls, piers and other permanent supports of buildings and structures shall be protected from frost by one or more of the following methods:

1. Extending below the frost line of the locality;
2. Constructing in accordance with ASCE-32; or
3. Erecting on solid rock.

Exception: Free-standing buildings meeting all of the following conditions shall not be required to be protected:

1. Classified in Occupancy Category I, in accordance with Section1604.5;
2. Area of 600 square feet (56 m^2) or less for light-frame construction or 400 square feet (37 m^2) or less for other than light-frame construction; and
3. Eave height of 10 feet (3048 mm) or less.

Footings shall not bear on frozen soil unless such frozen condition is of a permanent character.

Q: Are precast concrete stairs that are used primarily for one- or two-family dwellings required to be supported by footings extending below the frost line?

A: No. Section 1805.2.1 requires frost protection of the foundation systems only for buildings and structures that are larger than 600 square feet for light-frame construction or 400 square feet in area for other than light-frame construction or more than 10 feet in eave height. The amount of differential settlement caused by soil failure or frost action must fall within the exterior stairway allowance to abate the hazards to the individual using the stairs.

1805.5 Foundation walls. Concrete and masonry foundation walls shall be designed in accordance with Chapter 19 or 21, respectively. Foundation walls that are laterally supported at the top and bottom and within the parameters of Tables 1805.5(1) through 1805.5(5) are permitted to be designed and constructed in accordance with Sections 1805.5.1 through 1805.5.5.

Q: Table 1805.5(1) requires an 8-inch minimum thickness for foundation walls constructed of plain masonry. Does a foundation wall composed of 6-inch-thick hollow concrete block and 4-inch-thick solid brick masonry comply with the thickness requirements?

A: Yes. Such a foundation wall can be considered a 10-inch-thick composite foundation wall and would meet the minimum thickness requirements of Section 1805.5.1.2, provided that the wall is designed and constructed in accordance with ACI 530/ASCE 5/TMS 402 as a composite assembly that will adequately support the required vertical and lateral loads.

TABLE 1805.5(1), see page 85.

SECTION 1806 RETAINING WALLS

1806.1 General. Retaining walls shall be designed to ensure stability against overturning, sliding, excessive foundation pressure and water uplift. Retaining walls shall be designed for a safety factor of 1.5 against lateral sliding and overturning.

Q: Retaining walls must be designed to resist lateral soil loads, and the design must include a safety factor of 1.5 against lateral sliding and overturning. Is the design safety factor of 1.5 required to be applied to retaining walls for seismic loads?

A: No. Section 1806 establishes the minimum design requirements for basement walls and retaining walls with regards to lateral soil loads. Retaining walls subject to seismic loads must comply with seismic design provisions addressed elsewhere in the referenced *International Building Code.*

TABLE 1805.5(1)
PLAIN MASONRY FOUNDATION WALLS[a, b, c]

MAXIMUM WALL HEIGHT (feet)	MAXIMUM UNBALANCED BACKFILL HEIGHT[e] (feet)	MINIMUM NOMINAL WALL THICKNESS (inches)		
		Soil classes and lateral soil load[a] (psf per foot below natural grade)		
		GW, GP, SW and SP soils 30	GM, GC, SM, SM-SC and ML soils 45	SC, ML-CL and Inorganic CL soils 60
7	4 (or less)	8	8	8
	5	8	10	10
	6	10	12	10 (solid[c])
	7	12	10 (solid[c])	10 (solid[c])
8	4 (or less)	8	8	8
	5	8	10	12
	6	10	12	12 (solid[c])
	7	12	12 (solid[c])	Note d
	8	10 (solid[c])	12 (solid[c])	Note d
9	4 (or less)	8	8	8
	5	8	10	12
	6	12	12	12 (solid[c])
	7	12 (solid[c])	12 (solid[c])	Note d
	8	12 (solid[c])	Note d	Note d
	9	Note d	Note d	Note d

For SI: 1 inch = 25.4 mm, 1 foot = 304.8 mm, 1 pound per square foot per foot = 0.157kPa/m.

a. For design lateral soil loads, see Section 1610. Soil classes are in accordance with the Unified Soil Classification System and design lateral soil loads are for moist soil conditions without hydrostatic pressure.

b. Provisions for this table are based on construction requirements specified in Section 1805.5.2.2.

c. Solid grouted hollow units or solid masonry units.

d. A design in compliance with Chapter 21 or reinforcement in accordance with Table 1805.5(2) is required.

e. For height of unbalanced backfill, see Section 1805.5.1.2.

CHAPTER 23

WOOD

SECTION 2303
MINIMUM STANDARDS AND QUALITY

2303.1.4 Wood structural panels. Wood structural panels, when used structurally (including those used for siding, roof and wall sheathing, subflooring, diaphragms and built-up members), shall conform to the requirements for their type in DOC PS 1 or PS 2. Each panel or member shall be identified for grade and glue type by the trademarks of an approved testing and grading agency. Wood structural panel components shall be designed and fabricated in accordance with the applicable standards listed in Section 2306.1 and identified by the trademarks of an approved testing and inspection agency indicating conformance with the applicable standard. In addition, wood structural panels when permanently exposed in outdoor applications shall be of exterior type, except that wood structural panel roof sheathing exposed to the outdoors on the underside is permitted to be interior type bonded with exterior glue, Exposure 1.

Q: Are wood structural panels, such as plywood or oriented strand board (OSB), as described in Section 2303.1.4 of the *International Building Code*, required to be identified with a grade mark when used for a nonstructural application such as wall sheathing or exterior siding?

A: Yes. The provisions stated in Section 2303.1.4 only exempts plywood and/or OSB used in a nonstructural application from complying with DOC PS 1 or PS 2. All other requirements of Section 2303.1.4 apply whether the plywood or OSB is used structurally or nonstructurally.

2303.2.2 Strength adjustments. Design values for untreated lumber and wood structural panels, as specified in Section 2303.1, shall be adjusted for fire-retardant-treated wood. Adjustments to design values shall be based on an approved method of investigation that takes into consideration the effects of the anticipated temperature and humidity to which the fire-retardant-treated wood will be subjected, the type of treatment and redrying procedures.

Q: Is fire-retardant-treated wood required to be identified with information relating to the structural properties that occur following the fire-retardant treatment process?

A: Yes. Sections 2303.1.1 and 2303.1.4 require that lumber and plywood be marked in a manner indicating the structural properties of the untreated material. Section 2303.2.1 requires fire-retardant-treated lumber and plywood to bear the label of an approved agency and that such label indicates compliance with the testing standard and strength modifications necessary for lumber and plywood that are pressure impregnated with fire-retardant chemicals. This label is an important component in the evaluation of the effect that specific fire-retardant treatments have on the structural properties of lumber and plywood, and is required by both the code and the applicable referenced standards.

It is essential that the code official be provided with information enabling him or her to determine whether the product complies with the code and the industry standards referenced therein. The label required on the product must contain sufficient information to accomplish such purpose.

SECTION 2304
GENERAL CONSTRUCTION REQUIREMENTS

2304.6 Wall sheathing. Except as provided for in Section 1405 for weatherboarding or where stucco construction that complies with Section 2510 is installed, enclosed buildings shall be sheathed with one of the materials of the nominal thickness specified in Table 2304.6 or any other approved material of equivalent strength or durability.

Q: Are all wood frame buildings required to have exterior wall sheathing?

A: Yes. Sheathing is a structural element required by Section 2304.6. Siding is a material used on the exterior of the building to provide weather resistance as required by Section 1405.2. It is possible for a single material to provide structural and weather-resistant properties as indicated in Section 1405.2.

2304.6.1 Wood structural panel sheathing. Where wood structural panel sheathing is used as the exposed finish on the exterior of outside walls, it shall have an exterior exposure durability classification. Where wood structural panel sheathing is used on the exterior of outside walls but not as the exposed finish, it shall be of a type manufactured with exterior glue (Exposure 1 or Exterior). Where wood structural panel sheathing is used elsewhere, it shall be of a type manufactured with intermediate or exterior glue.

Q: Are wood structural panels that are only used as a sheathing material required to be exterior grade?

A: No. Section 2304.6.1 requires wood structural panels for exterior installation to be of the exterior-waterproof type. This applies to wood structural panels intended to be left exposed to the elements. Wood structural panels that will be protected by an approved roof or wall covering would not be considered an "exterior installation" and, therefore, are not required to be of the "exterior-waterproof type."

2304.11.2.2 Wood suported by exterior foundation. Wood framing members, including wood sheathing, that rest on exterior foundation walls and are less than 8 inches (203 mm) from exposed earth shall be of naturally durable or preservative-treated wood.

Q: Does the requirement for all wood framing members to be 8 inches above exposed earth in Section 2304.11.2.2 apply to buildings or structures classified as Group U (such as detached private garages and storage sheds)?

A: Yes. Section 2304.11.2.2 sets forth this requirement as applicable to all buildings. Section 312.1 requires that buildings and structures of an accessory character, including detached private garages and storage sheds, be constructed commensurate with the fire and life hazard incidental to their occupancy. Because structural performance is an essential building safety component, compliance with Section 2304.11.2.2 is required for these types of occupancies.

2304.11.2.4 Sleepers and sills. Sleepers and sills on a concrete or masonry slab that is in direct contact with earth shall be of naturally durable or preservative-treated wood.

Q: Section 2304.11.2.4 requires sleepers and sills located on a concrete or masonry slab that is in direct contact with earth to be of approved naturally durable or preservative-treated wood. If the slab is protected by a moisture proof membrane, does that mean the slab is not in direct contact with earth and, therefore, preservative-treated wood is not necessary?

A: Yes. Section 2304.11.2.4 requires the use of decay- and termite-resistant wood (naturally durable or preservative treated) whenever located on slabs that are in direct contact with earth. An approved membrane located between the slab and the earth eliminates direct contact. As a result, the use of decay- and termite-resistant wood is not required.

CHAPTER 24

GLASS AND GLAZING

SECTION 2405
SLOPED GLAZING AND SKYLIGHTS

2405.3 Screening. Where used in monolithic glazing systems, heat-strengthened glass and fully tempered glass shall have screens installed below the glazing material. The screens and their fastenings shall: (1) be capable of supporting twice the weight of the glazing; (2) be firmly and substantially fastened to the framing members; and (3) be installed within 4 inches (102 mm) of the glass. The screens shall be constructed of a noncombustible material not thinner than No. 12 B&S gage (0.0808 inch) with mesh not larger than 1 inch by 1 inch (25 mm by 25 mm). In a corrosive atmosphere, structurally equivalent noncorrosive screen materials shall be used. Heat-strengthened glass, fully tempered glass and wired glass, when used in multiple-layer glazing systems as the bottom glass layer over the walking surface, shall be equipped with screening that conforms to the requirements for monolithic glazing systems.

Exception: In monolithic and multiple-layer sloped glazing systems, the following applies:

1. Fully tempered glass installed without protective screens where glazed between intervening floors at a slope of 30 degrees (0.52 rad) or less from the vertical plane shall have the highest point of the glass 10 feet (3048 mm) or less above the walking surface.
2. Screens are not required below any glazing material, including annealed glass, where the walking surface below the glazing material is permanently protected from the risk of falling glass or the area below the glazing material is not a walking surface.
3. Any glazing material, including annealed glass, is permitted to be installed without screens in the sloped glazing systems of commercial or detached noncombustible greenhouses used exclusively for growing plants and not open to the public, provided that the height of the greenhouse at the ridge does not exceed 30 feet (9144 mm) above grade.
4. Screens shall not be required within individual dwelling units in Groups R-2, R-3 and R-4 as applicable in Section 101.2 where fully tempered glass is used as single glazing or as both panes in an insulating glass unit, and the following conditions are met:
    - 4.1. Each pane of the glass is 16 square feet (1.5 m^2) or less in area.
    - 4.2. The highest point of the glass is 12 feet (3658 mm) or less above any walking surface or other accessible area.
    - 4.3. The glass thickness is $^3/_{16}$ inch (4.8 mm) or less.
5. Screens shall not be required for laminated glass with a 15-mil (0.38 mm) polyvinyl butyral (or equivalent) interlayer used within individual dwelling units in Groups R-2, R-3 and R-4 as applicable in Section 101.2 within the following limits:
    - 5.1. Each pane of glass is 16 square feet (1.5 m^2) or less in area.
    - 5.2. The highest point of the glass is 12 feet (3658 mm) or less above a walking surface or other accessible area.

Q: Where used as the bottom layer in multiple-layer glazing systems, is laminated glass required to be equipped with screening or other permanent protection whenever installed over walking surfaces?

A: No. In general, Section 2405.3 does not require screens under laminated glass because of the tendency of laminated glass to hold together as a unit upon breaking. Furthermore, Section 2405.3 only requires glazing systems with bottom layers of heat-strengthened, fully tempered and wired glass to be provided with screens or other permanent protection. Breakage of glass layers above a layer of laminated glass will be prevented from falling to the surface below by the bottom layer of laminated glass. Therefore, screens are not required to be provided whenever laminated glass is the bottom layer in glazing systems located above walking surfaces.

2405.5 Unit skylights. Unit skylights shall be tested and labeled as complying with AAMA/WDMA/CSA 101/I.S.2/A440. The label shall state the name of the manufacturer, the approved labeling agency, the product designation and the performance grade rating as specified in AAMA/WDMA/CSA 101/I.S.2/A440. If the product manufacturer has chosen to have the performance grade of the skylight rated separately for positive and negative design pressure, then the label shall state both performance grade ratings as specified in AAMA/WDMA/CSA 101/I.S.2/A440 and the skylight shall comply with Section 2405.5.2. If the skylight is not rated separately for positive and negative pressure, then the performance grade rating shown on the label shall be the performance grade rating determined in accordance with AAMA/WDMA/CSA 101/I.S.2/A440 for both positive and negative design pressure and the skylight shall conform to Section 2405.5.1.

Q: Are unit skylights as described in Section 2405.5 of the *International Building Code* required to comply with all the requirements of the referenced standard applicable to unit skylights?

A: Yes. Where compliance to any specifically identified portion(s) of any standard referenced in the *International Codes* is not provided in the text of the code, compliance with all applicable provisions in the referenced standard is required.

SECTION 2406 SAFETY GLAZING

2406.3 Hazardous locations. The following shall be considered specific hazardous locations requiring safety glazing materials:

1. Glazing in swinging doors except jalousies (see Section 2406.3.1).
2. Glazing in fixed and sliding panels of sliding door assemblies and panels in sliding and bifold closet door assemblies.
3. Glazing in storm doors.
4. Glazing in unframed swinging doors.
5. Glazing in doors and enclosures for hot tubs, whirlpools, saunas, steam rooms, bathtubs and showers. Glazing in any portion of a building wall enclosing these compartments where the bottom exposed edge of the glazing is less than 60 inches (1524 mm) above a standing surface.
6. Glazing in an individual fixed or operable panel adjacent to a door where the nearest exposed edge of the glazing is within a 24-inch (610 mm) arc of either vertical edge of the door in a closed position and where the bottom exposed edge of the glazing is less than 60 inches (1524 mm) above the walking surface.

   Exceptions:
   1. Panels where there is an intervening wall or other permanent barrier between the door and glazing.
   2. Where access through the door is to a closet or storage area 3 feet (914 mm) or less in depth. Glazing in this application shall comply with Section 2406.3, Item 7.
   3. Glazing in walls perpendicular to the plane of the door in a closed position, other than the wall towards which the door swings when opened, in one- and two-family dwellings or within dwelling units in Group R-2.
7. Glazing in an individual fixed or operable panel, other than in those locations described in preceding Items 5 and 6, which meets all of the following conditions:
   - 7.1. Exposed area of an individual pane greater than 9 square feet (0.84 m^2);
   - 7.2. Exposed bottom edge less than 18 inches (457 mm) above the floor;
   - 7.3. Exposed top edge greater than 36 inches (914 mm) above the floor; and
   - 7.4. One or more walking surface(s) within 36 inches (914 mm) horizontally of the plane of the glazing.

   Exception: Safety glazing for Item 7 is not required for the following installations:
   1. A protective bar $1^1/_2$ inches (38 mm) or more in height, capable of withstanding a horizontal load of 50 pounds plf (730 N/m) without contacting the glass, is installed on the accessible sides of the glazing 34 inches to 38 inches (864 mm to 965 mm) above the floor.
   2. The outboard pane in insulating glass units or multiple glazing where the bottom exposed edge of the glass is 25 feet (7620 mm) or more above any grade, roof, walking surface or other horizontal or sloped (within 45 degrees of horizontal) (0.78 rad) surface adjacent to the glass exterior.
8. Glazing in guards and railings, including structural baluster panels and nonstructural in-fill panels, regardless of area or height above a walking surface.
9. Glazing in walls and fences enclosing indoor and outdoor swimming pools, hot tubs and spas where all of the following conditions are present:
   - 9.1. The bottom edge of the glazing on the pool or spa side is less than 60 inches (1524 mm) above a walking surface on the pool or spa side of the glazing; and
   - 9.2. The glazing is within 60 inches (1524 mm) horizontally of the water's edge of a swimming pool or spa.
10. Glazing adjacent to stairways, landings and ramps within 36 inches (914 mm) horizontally of a walking surface; when the exposed surface of the glass is less than 60 inches (1524 mm) above the plane of the adjacent walking surface.
11. Glazing adjacent to stairways within 60 inches (1524 mm) horizontally of the bottom tread of a stairway in any direction when the exposed surface of the glass is less than 60 inches (1524 mm) above the nose of the tread.

    Exception: Safety glazing for Item 10 or 11 is not required for the following installations where:
    1. The side of a stairway, landing or ramp which has a guardrail or handrail, including balusters or in-fill panels, complying with the provisions of Sections 1012 and 1607.7; and
    2. The plane of the glass is greater than 18 inches (457 mm) from the railing.

Q: Does the requirement for safety glazing in Item 6 of Section 2406.3 apply to all buildings and to all types of glazed openings, including double-hung windows, picture windows and standard door sidelights that fall within the specified proximity to the door?

A: Yes. The requirement for safety glazing adjacent to ingress and egress doors is intended to prevent injury due to accidental impact with the glazing that may occur through use of the door. The need for this protection is equally applicable to all buildings, regardless of use, and to all types of glazed openings within the specified parameters and exceptions set forth in Item 6 of Section 2406.3.

Q: Where multiple-layer glazing is used in locations required by Section 2406.3 to be of safety glazing, is safety glazing required for only the pane or panes facing the side from which impact will occur?

A: No. It cannot be assumed that a person impacting a glazed panel will stop at the first pane of safety glazing and not be injured by the other pane or panes. Therefore, except for the specific locations identified in Exception 7, all panes in multiple-layer systems are required to be composed of safety glazing. The panes of glass, however, need not consist of identical material as long as they contain approved safety glazing materials.

Q: Where a wall opening that is greater than 9 square feet in total area and located less than 18 inches above an adjacent floor level is glazed with a single fixed panel of glass, Item 7 of Section 2406.3 requires the glass to be of an approved safety glazing material. Would safety glazing be required if the opening were glazed with individually framed fixed-glass panels; each of which is less than 9 square feet in area?

A: No. The intent of Item 7 of Section 2406.3 is to require safety glazing wherever an individual panel of fixed glass is greater than 9 square feet in area and located relative to the floor as specified. Individual glass panels that are 9 square feet or less in area have not posed the same risk of injury upon breakage as have larger glass panels. Furthermore, such an opening with smaller, individually framed panels is more likely to be noticeable and, therefore, less likely to be subjected to accidental impact.

Q: Is it the intent of Item 7 of Section 2406.3 that safety glazing be provided in fixed or operable panels?

A: Yes. The code specifically states "fixed or operable panel." A "fixed" panel is one that, when subject to human impact, will not swing or release freely, thus not allowing a person to make contact with the panel without possibly breaking the glass. The potential hazard presented by a glazed panel of the size and location described in Item 7 of Section 2406.3 that is operable but does not swing or release freely upon contact is the same as the potential hazard of a glazed panel that is nonoperable.

Q: Is it the intent of Item 7 of Section 2406.3 to require safety glazing in glazed panels that comply with all of the listed criteria; that is, have an individual pane exceeding 9 square feet, the exposed bottom edge located less than 18 inches above the floor, the exposed top edge greater than 36 inches above the floor and a walking surface falling within 36 inches of the panels?

A: Yes. It is the intent of the code that a glazed panel must satisfy all of the criteria listed in Item 7 of Section 2406.3 before safety glazing is required.

CHAPTER 26

PLASTIC

SECTION 2603
FOAM PLASTIC INSULATION

2603.3 Surface-burning characteristics. Unless otherwise indicated in this section, foam plastic insulation and foam plastic cores of manufactured assemblies shall have a flame spread index of not more than 75 and a smoke-developed index of not more than 450 where tested in the maximum thickness intended for use in accordance with ASTM E 84. Loose fill-type foam plastic insulation shall be tested as board stock for the flame spread index and smoke-developed index.

Exceptions:

1. Smoke-developed index for interior trim as provided for in Section 2604.2.
2. In cold storage buildings, ice plants, food plants, food processing rooms and similar areas, foam plastic insulation where tested in a thickness of 4 inches (102 mm) shall be permitted in a thickness up to 10 inches (254 mm) where the building is equipped throughout with an automatic fire sprinkler system in accordance with Section 903.3.1.1. The approved automatic sprinkler system shall be provided in both the room and that part of the building in which the room is located.
3. Foam plastic insulation that is a part of a Class A, B or C roof-covering assembly provided the assembly with the foam plastic insulation satisfactorily passes FM 4450 or UL 1256. The smoke-developed index shall not be limited for roof applications.
4. Foam plastic insulation greater than 4 inches (102 mm) in thickness shall have a maximum flame spread index of 75 and a smoke-developed index of 450 where tested at a minimum thickness of 4 inches (102 mm), provided the end use is approved in accordance with Section 2603.8 using the thickness and density intended for use.
5. Flame spread and smoke-developed indexes for foam plastic interior signs in covered mall buildings provided the signs comply with Section 402.14.

Q: During the ASTM E 84 test of some thermoplastic foam plastic materials, the material softens and falls from the installed position to the floor of the test furnace. It is common for the testing laboratory to report the burning of the material on the furnace floor as a footnote to the flame spread and smoke-developed ratings. Is it the intent of Section 2603.3 to use the footnoted value of the smoke-developed rating from the test report in determining compliance with the code?

A: No. Although the footnoted value would indicate the smoke generated for the entire 10-minute test period, the ASTM E 84 test method is not intended to deal with the contribution of the test materials that fall from the ceiling test position. For this reason, the smoke-developed rating is reported with a footnote in accordance with the formal test procedure. In referencing ASTM E 84, the code is anticipating a test performance in compliance with the formal procedure, with any attributes and defects associated with such test method.

Q: Are scrap expanded polystyrene beads, which are similar to those used as packing material, permitted for use as masonry cavity wall insulation?

A: No. Only foam plastic insulation that is properly labeled and has been determined to have a flame spread rating of not more than 75 and a smoke-developed rating of not more than 450, when tested in accordance with ASTM E 84, shall be installed as masonry cavity wall insulation in accordance with Section 2603.4.1.1. The use of the term "scrap" means that no such information is available.

CHAPTER 27

ELECTRICAL

SECTION 2701
GENERAL

2701.1 Scope. This chapter governs the electrical components, equipment and systems used in buildings and structures covered by this code. Electrical components, equipment and systems shall be designed and constructed in accordance with the provisions of the ICC *Electrical Code.*

Q: Is electrical equipment that is not part of a building's service equipment/system regulated by Section 605.7 of the *International Fire Code* and Section 2701.1 of the *International Building Code*?

A: No. Section 2701.1 states that the provisions of the code shall control the design and construction of all new installations of electrical conductors, equipment and systems in buildings or structures. Section 605.7 of the *International Fire Code* requires electrical appliances and fixtures to be tested and listed. Only equipment and systems that are used as the building's electrical service and distribution system are intended to be controlled by the *International Codes.* All electrically controlled devices that are specifically designed and necessary for the general safety and welfare of the occupants, as well as other equipment and systems that are specifically addressed in the referenced standard, NFPA 70, would need to comply with the code provisions. The intended scope of Section 601.1 of the *International Fire Code* and Section 2701.1 of the *International Building Code* is the same as the intended scope of NFPA 70.

CHAPTER 30

ELEVATORS AND CONVEYING SYSTEMS

SECTION 3005
CONVEYING SYSTEMS

3005.1 General. Escalators, moving walks, conveyors, personnel hoists and material hoists shall comply with the provisions of this section.

Q: Is a permit required for the installation of a conveyor system?

A: Yes. Conveyors and special hoisting and conveying equipment referenced in Section 3001.1 are within the scope of the code. Section 3005 contains specific requirements governing conveyors. Therefore, a permit is required in accordance with Section 105.1.

CHAPTER 31

SPECIAL CONSTRUCTION

SECTION 3102
MEMBRANE STRUCTURES

3102.1 General. The provisions of this section shall apply to air-supported, air-inflated, membrane-covered cable and membrane-covered frame structures, collectively known as membrane structures, erected for a period of 180 days or longer. Those erected for a shorter period of time shall comply with the *International Fire Code*. Membrane structures covering water storage facilities, water clarifiers, water treatment plants, sewage treatment plants, greenhouses and similar facilities not used for human occupancy, are required to meet only the requirements of Sections 3102.3.1 and 3102.7.

Q: Is Section 3102.1 intended to apply to tents only for assembly purposes?

A: No.

Q: If so, would this apply only to an occupant load of more than 50 in accordance with Section 303.1, Exception 2?

A: No. Section 3102.1 applies to tents, membrane structures and other temporary structures erected for a period of 180 days or longer. This section makes no reference to specific occupancies. Permits are required whenever the temporary structure exceeds the 120-square-foot area limitation and is used for 10 or more people. The requirements specified in Section 3102 are equally necessary and applicable to all occupancies.

SECTION 3103
TEMPORARY STRUCTURES

3103.1.1 Permit required. Temporary structures that cover an area in excess of 120 square feet (11.16 m^2), including connecting areas or spaces with a common means of egress or entrance which are used or intended to be used for the gathering together of ten or more persons, shall not be erected, operated or maintained for any purpose without obtaining a permit from the building official.

Q: Can a temporary structure be placed on a permanent foundation?

A: Yes.

Q: For what length of time can a structure remain in a particular location and still be considered of a temporary nature?

A: The length of time would be 180 days by permit from the code official or one year through a variance granted by the local board of appeals. Section 3103.1.1 requires a special permit for buildings or structures of a temporary nature. This permit can be granted for a period of less than 180 days. The building official is authorized to grant extensions for demonstrated cause in accordance with Section 107.1.

SECTION 3109
SWIMMING POOL ENCLOSURES AND SAFETY DEVICES

3109.4 Residential swimming pools. Residential swimming pools shall comply with Sections 3109.4.1 through 3109.4.3.

> **Exception:** A swimming pool with a power safety cover or a spa with a safety cover complying with ASTM F 1346.

3109.4.1 Barrier height and clearances. The top of the barrier shall be at least 48 inches (1219 mm) above grade measured on the side of the barrier that faces away from the swimming pool. The maximum vertical clearance between grade and the bottom of the barrier shall be 2 inches (51 mm) measured on the side of the barrier that faces away from the swimming pool. Where the top of the pool structure is above grade, the barrier is authorized to be at ground level or mounted on top of the pool structure, and the maximum vertical clearance between the top of the pool structure and the bottom of the barrier shall be 4 inches (102 mm).

Q: A private swimming pool is erected above ground with the pool walls extending at least 4 feet above the surrounding finished ground level. The pool is equipped with an elevated walkway located along the upper edge of the pool walls, and access from the ground level to the walkway is provided by a permanent nonretractable ladder. The property on which the pool is erected is not fenced. Does the pool comply with Section 3109.4.1 requirements?

A: No. The intent of Section 3109.4.1 is to provide a continuous barrier at least 4 feet in height, making the pool inaccessible to small children without adult intervention. The height of the pool located above the ground provides the necessary barrier, except for the ladder. Therefore, the ladder would have to be surrounded by a fence or other barrier in accordance with Section 3109.4.1.9.

3109.4.1.8 Dwelling wall as a barrier. Where a wall of a dwelling serves as part of the barrier, one of the following shall apply:

1. Doors with direct access to the pool through that wall shall be equipped with an alarm that produces an audible warning when the door and/or its screen, if present, are opened. The alarm shall be listed in accordance with UL 2017. The audible alarm shall activate within 7 seconds and sound continuously for a minimum of 30 seconds after the door and/or its screen, if present, are opened and be capable of being heard throughout the house during normal household activities. The alarm shall automatically reset under all conditions. The alarm shall be equipped with a manual means, such as touchpad or switch, to temporarily deactivate the alarm for a single opening. Such deactivation shall last for not more than 15 seconds. In dwellings not required to be Accessible, Type A or Type B units, the deactivation switch shall be located 54 inches (1372 mm) or more above the threshold of the door. In dwellings required to be Accessible, Type A or Type B units, the deactivation switch(es) shall be located at 54 inches (1372 mm) maximum and 48 inches minimum above the threshold of the door.
2. The pool shall be equipped with a power safety cover that complies with ASTM F 1346.
3. Other means of protection, such as self-closing doors with self-latching devices, which are approved by the administrative authority, shall be accepted so long as the degree of protection afforded is not less than the protection afforded by Section 3109.4.1.8, Item 1 or 2.

Q: Is the audible pool alarm device required by Section 3109.4.1.8(1) to be connected to an AC primary power source with a battery backup system?

A: No. Section 3109.4.1.8(1) does not specify the type of power source required for the door alarm. Furthermore, the code does not require that battery backup be provided.

Q: Where the wall of a dwelling unit serves as part of the barrier to a residential swimming pool, are the required door alarms permitted to be battery operated?

A: Yes. The provisions for barrier alarms in Item 1 of Section 3109.4.1.8 establishes when and where barrier alarms are required, as well as how the alarms must perform. The provisions do not stipulate a specific power source for the barrier alarms. The owner of the dwelling or the owner's agent must provide sufficient documentation to verify that the electrical power source provided to the alarms will enable them to perform as required.

Q: Is the activation device required to be interconnected to additional alarms located on each floor (including basements) of the dwelling?

A: No. Section 3109.4.1.8(1) requires that the door alarm emit an audible warning. The minimum sound pressure level rating from a single source is viewed as providing an adequate warning to a supervising adult within the dwelling. It should be noted that the code requires the alarm to be capable of being heard throughout the house during normal household activities.

CHAPTER 33

SAFEGUARDS DURING CONSTRUCTION

SECTION 3301
GENERAL

3301.1 Scope. Provisions of this chapter shall govern safety during construction and the protection of adjacent public and private properties.

Q: Do the requirements of Chapter 33 specifically apply to projects such as the widening or deepening of an existing river for flood control purposes or any other similar project?

A: No. Section 3301.1 states that these provisions govern safety during construction. Section 101.2 establishes the scope of the code as applicable to buildings or structures. The widening of a river does not meet the definition of a building or structure in Section 202 and, therefore, the provisions of Chapter 33 are not applicable.

CHAPTER 34

EXISTING STRUCTURES

SECTION 3401
GENERAL

3401.2 Maintenance. Buildings and structures, and parts thereof, shall be maintained in a safe and sanitary condition. Devices or safeguards which are required by this code shall be maintained in conformance with the code edition under which installed. The owner or the owner's designated agent shall be responsible for the maintenance of buildings and structures. To determine compliance with this subsection, the building official shall have the authority to require a building or structure to be reinspected. The requirements of this chapter shall not provide the basis for removal or abrogation of fire protection and safety systems and devices in existing structures.

Q: What is the definition of "Maintenance" as interpreted and enforced under Section 3401.2?

A: "Maintenance" means to preserve or keep in proper condition. The building official has the authority to rule on the performance of maintenance work whenever public health or safety is affected. Under Section 3401.3, the building official has the authority to require a building to be maintained in accordance with the requirements of the *International Codes.*

Q: Is it the intent of Section 3401.2 that a required fire protection system, such as an automatic sprinkler system, be maintained even though it is not required for new construction?

A: Yes. Section 3401.2 specifically calls for maintenance of such equipment or devices required by the current code. This section further states that the requirements of Chapter 34 "shall not provide the basis for removal or abrogation of fire protection and safety systems and devices in existing structures."

SECTION 3403
ADDITIONS, ALTERATIONS OR REPAIRS

3403.1 Existing buildings or structures. Additions or alterations to any building or structure shall conform with the requirements of the code for new construction. Additions or alterations shall not be made to an existing building or structure which will cause the existing building or structure to be in violation of any provisions of this code. An existing building plus additions shall comply with the height and area provisions of Chapter 5. Portions of the structure not altered and not affected by the alteration are not required to comply with the code requirements for a new structure.

Q: If an addition to an existing building is separated by a fire wall complying with all applicable requirements of Section 705, would the existing building be required to conform to the current code requirements regarding new construction based on Section 3403.1 for an increase in height and area?

A: No. Section 102.6 allows existing buildings to be maintained in a manner so they do not pose a hazard to occupants and the general public. By separating an addition from the existing building with a fire wall, such addition is considered a separate building and, therefore, the existing building has not been increased in size. Moreover, an addition to an existing building is allowed by Section 3403.1 if such addition does not further remove the existing building from its current degree of code compliance.

In this particular instance, such an addition is permitted if the existing building does not require the frontage increase in Section 506.2 for current code conformance. If the existing building does not conform to current height and area limitations (Sections 503, 504 and 506), the addition is not permitted, as it would further remove the existing building from its current degree of code compliance with height and area limitations.

Q: If an existing building is equipped throughout with a required automatic fire suppression system and is altered by installing walls and ceilings to create new interior rooms or spaces, can suppression be omitted from those new spaces if the enclosing walls and ceilings are of noncombustible or fire-resistance-rated construction?

A: No. Section 3403.1 requires all alterations to an existing structure to conform to code requirements for a new structure. Therefore, in a building required to be equipped throughout with an automatic fire suppression system, suppression must be installed in all spaces as required by the applicable referenced standards for the type of suppression system. There are no provisions for buildings required to be equipped throughout with a suppression system that permits the omission of suppression from a space on the basis of noncombustible or fire-resistance-rated enclosure walls. Furthermore, the existing suppression system must conform or be altered to conform to the required spacing and location provisions for the new room layout and arrangement.

Q: A manual fire alarm system is being replaced in an existing Group I-2 occupancy. There are patient rooms that are not currently protected by smoke detectors or quick-response sprinklers as required by Section 407.5. Are either the system requirements of Section 407.5 applicable or additional smoke detectors required to be provided in the patient rooms by virtue of the manual fire alarm system being replaced?

A: No. It is assumed that prior to the proposed replacement of the manual fire alarm system, the building has been approved as an existing building. In general, Section 3403.1 requires that all portions of the alteration must comply with the current code requirements. Those portions that are not being altered are not required to be brought up to code due to the proposed alteration. In this case, those portions of the fire alarm system being replaced must be in compliance with the current code. The installation of either smoke detectors or quick-response sprinkler heads in patient sleeping rooms (see Section 407.5) is not within the scope of the alteration or "affected by the alteration" as stipulated in Section 3403.1.

SECTION 3406
CHANGE OF OCCUPANCY

3406.1 Conformance. No change shall be made in the use or occupancy of any building that would place the building in a different division of the same group of occupancy or in a different group of occupancies, unless such building is made to comply with the requirements of this code for such division or group of occupancy. Subject to the approval of the building official, the use or occupancy of existing buildings shall be permitted to be changed and the building is allowed to be occupied for purposes in other groups without conforming to all the requirements of this code for those groups, provided the new or proposed use is less hazardous, based on life and fire risk, than the existing use.

Q: Does Section 3406.1 apply only to changes in occupancy involving special uses and occupancies identified in Chapter 4?

A: No. The intent of Section 3406.1 is that the code official be notified of and act to approve or deny any change in occupancy. Change of occupancy is a change in the purpose or level of activity within a structure that involves a change in application of the requirements of this code. The extent of code compliance for the existing building involved in the change in occupancy depends on the code official's determination of the intent of the provisions of law for the new use and occupancy. The primary consideration is to determine that no greater hazard to public safety or welfare results because of a change in occupancy.

SECTION 3407
HISTORIC BUILDINGS

3407.1 Historic buildings. The provisions of this code relating to the construction, repair, alteration, addition, restoration and movement of structures, and change of occupancy shall not be mandatory for historic buildings where such buildings are judged by the building official to not constitute a distinct life safety hazard.

Q: Is it the intent of Section 3407.1 to waive the fire protection system requirements whenever a building is classified as a "historic building"?

A: No. Section 3407.1 establishes that the code provisions are not mandatory for existing historical buildings only when such a modification is judged by the building official ". . . to not constitute a distinct life safety hazard." Simply stated, the intent of this section is to ensure that any conflicts are minimized between historical preservation and life safety. However, the overriding concern is life safety, and it is the intent of this section to ensure that such provisions of the code are met.

APPENDIX D

FIRE DISTRICTS

SECTION D102
BUILDING RESTRICTIONS

D102.1 Types of construction permitted. Within the fire district every building hereafter erected shall be either Type I, II, III or IV, except as permitted in Section D104.

D102.2.1 Exterior walls. Exterior walls of buildings located in the fire district shall comply with the requirements in Table 601 except as required in Section D102.2.6.

Q: Are metal buildings of Type II construction permitted inside the fire districts?

A: Yes. Section D102.1 permits Type II construction within the fire districts. Section D102.2.1 contains the requirements for fire resistance ratings of exterior walls.

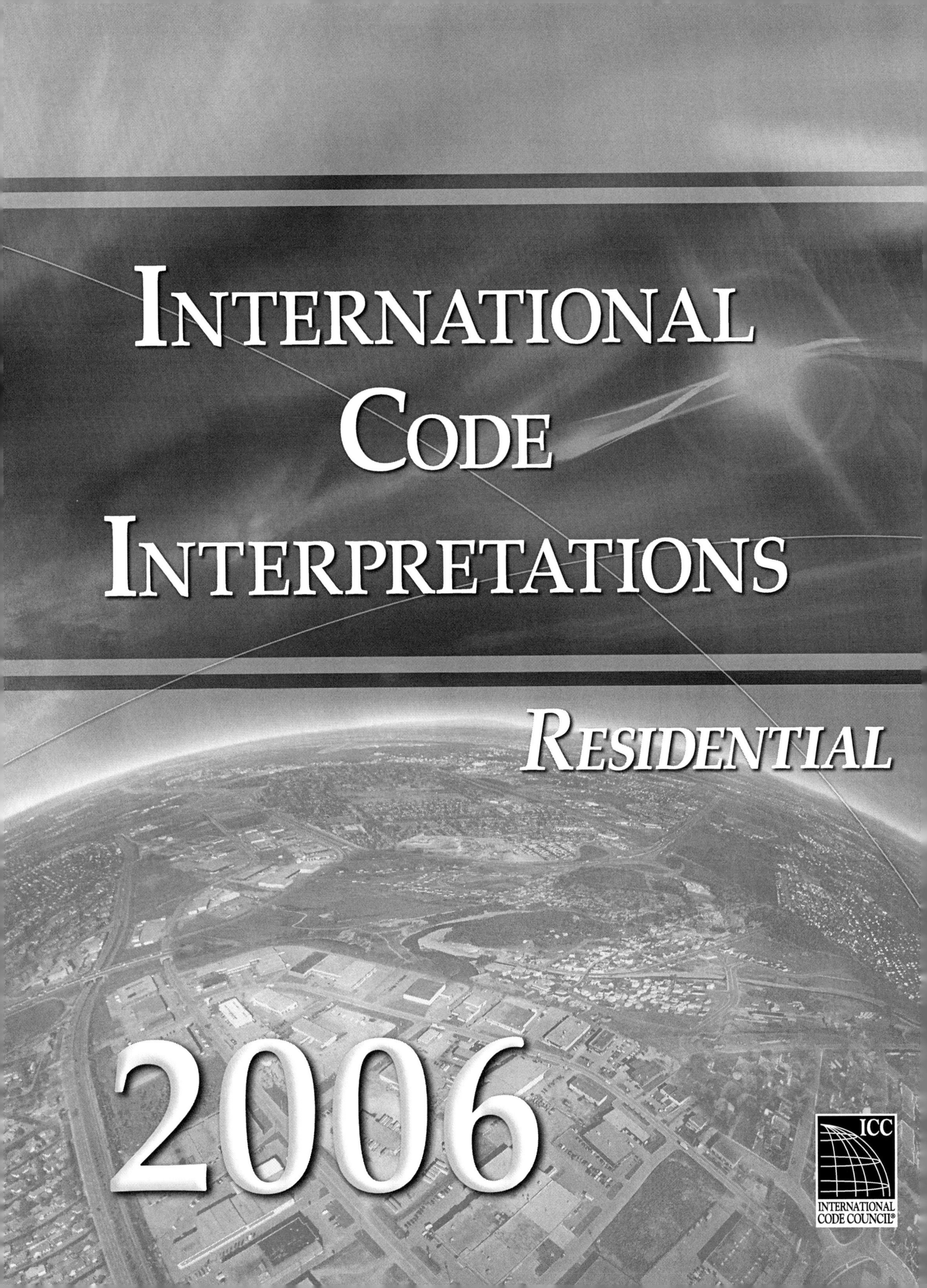
INTERNATIONAL
CODE
INTERPRETATIONS
RESIDENTIAL
2006
ICC
INTERNATIONAL
CODE COUNCIL®

CHAPTER 1

ADMINISTRATION

SECTION R105 PERMITS

R105.3.1.1 Determination of substantially improved or substantially damaged existing buildings in flood hazard areas. For applications for reconstruction, rehabilitation, addition or other improvement of existing buildings or structures located in an area prone to flooding as established by Table R301.2(1), the building official shall examine or cause to be examined the construction documents and shall prepare a finding with regard to the value of the proposed work. For buildings that have sustained damage of any origin, the value of the proposed work shall include the cost to repair the building or structure to its predamage condition. If the building official finds that the value of proposed work equals or exceeds 50 percent of the market value of the building or structure before the damage has occurred or the improvement is started, the finding shall be provided to the board of appeals for a determination of substantial improvement or substantial damage. Applications determined by the board of appeals to constitute substantial improvement or substantial damage shall meet the requirements of Section R324.

REFERENCED SECTIONS:

SECTION R102 APPLICABILITY

R102.7.1 Additions, alterations or repairs. Additions, alterations or repairs to any structure shall conform to that required for a new structure without requiring the existing structure to comply with all of the requirements of this code, unless otherwise stated. Additions, alterations or repairs shall not cause an existing structure to become unsafe or adversely affect the performance of the building.

TABLE R301.2(1) CLIMATIC AND GEOGRAPHIC DESIGN CRITERIA FLOOD HAZARDS[h]

h. The jurisdiction shall fill in this part of the table with (a) the date of the jurisdiction's entry into the National Flood Insurance Program (date of adoption of the first code or ordinance for management of flood hazard areas), (b) the date(s) of the currently effective FIRM and FBFM, or other flood hazard map adopted by the community, as may be amended.

Q: An existing dwelling constructed prior to the effective date established in Table R301.2(1) is located in an area prone to flooding. Under the applicable provisions of the 2006 *International Residential Code*, application is made for a permit to perform work on the existing dwelling. In accordance with Section R105.3.1.1, the building official determines the value of the proposed work to be less than 50 percent of the market value of the existing dwelling. Is the proposed work subject to the provisions of Section R323?

A: No. When proposed work is determined by the building official to be less than 50 percent of the market value of the existing dwelling, the application for permit is not required to be submitted for appeal, and the proposed work is not required to comply with the provisions of Section R323. In addition to the proposed work not being subject to the provisions of Section R323, the portions of the existing dwelling not affected by the proposed work is not required to be modified to comply with the provisions of Section R323. Although the provisions of Section R323 are deemed inapplicable to the existing dwelling and to the work proposed to the existing dwelling, that work which is permitted must comply with all other applicable provisions of the code as provided in Section R102.7.1.

Only when the building official determines the proposed work to be equal to or greater than 50 percent of the market value of the existing dwelling must the application for permit be presented for appeal.

CHAPTER 3

BUILDING PLANNING

SECTION R301
DESIGN CRITERIA

R301.5 Live load. The minimum uniformly distributed live load shall be as provided in Table R301.5.

TABLE R301.5
MINIMUM UNIFORMLY DISTRIBUTED LIVE LOADS
(in pounds per square foot)

USE	LIVE LOAD
Attics with limited storage[b, g, h]	20
Attics without storage[b]	10
Decks[e]	40
Exterior balconies	60
Fire escapes	40
Guardrails and handrails[d]	200[i]
Guardrails in–fill components[f]	50[i]
Passenger vehicle garages[a]	50[a]
Rooms other than sleeping rooms	40
Sleeping rooms	30
Stairs	40[c]

For SI: 1 pound per square foot = 0.0479 kPa, 1 square inch = 645 mm^2,
1 pound = 4.45 N.

a. Elevated garage floors shall be capable of supporting a 2,000-pound load applied over a 20-square-inch area.

b. Attics without storage are those where the maximum clear height between joist and rafter is less than 42 inches, or where there are not two or more adjacent trusses with the same web configuration capable of containing a rectangle 42 inches high by 2 feet wide, or greater, located within the plane of the truss. For attics without storage, this live load need not be assumed to act concurrently with any other live load requirements.

c. Individual stair treads shall be designed for the uniformly distributed live load or a 300-pound concentrated load acting over an area of 4 square inches, whichever produces the greater stresses.

d. A single concentrated load applied in any direction at any point along the top.

e. See Section R502.2.1 for decks attached to exterior walls.

f. Guard in-fill components (all those except the handrail), balusters and panel fillers shall be designed to withstand a horizontally applied normal load of 50 pounds on an area equal to 1 square foot. This load need not be assumed to act concurrently with any other live load requirement.

g. For attics with limited storage and constructed with trusses, this live load need be applied only to those portions of the bottom chord where there are two or more adjacent trusses with the same web configuration capable of containing a rectangle 42 inches high or greater by 2 feet wide or greater, located within the plane of the truss. The rectangle shall fit between the top of the bottom chord and the bottom of any other truss member, provided that each of the following criteria is met:

1. The attic area is accessible by a pull-down stairway or framed opening in accordance with Section R807.1; and
2. The truss has a bottom chord pitch less than 2:12.

h. Attic spaces served by a fixed stair shall be designed to support the minimum live load specified for sleeping rooms.

i. Glazing used in handrail assemblies and guards shall be designed with a safety factor of 4. The safety factor shall be applied to each of the concentrated loads applied to the top of the rail, and to the load on the in-fill components. These loads shall be determined independent of one another, and loads are assumed not to occur with any other live load.

Q: Is the 20 psf live load for attics with storage required to be applied concurrent with other required live loads?

A: Yes.

SECTION R302
EXTERIOR WALL LOCATION

R302.1 Exterior walls. Construction, projections, openings and penetrations of exterior walls of dwellings and accessory buildings shall comply with Table R302.1. These provisions shall not apply to walls, projections, openings or penetrations in walls that are perpendicular to the line used to determine the fire separation distance. Projections beyond the exterior wall shall not extend more than 12 inches (305 mm) into the areas where openings are prohibited.

Exceptions:

1. Detached tool sheds and storage sheds, playhouses and similar structures exempted from permits are not required to provide wall protection based on location on the lot. Projections beyond the exterior wall shall not extend over the lot line.
2. Detached garages accessory to a dwelling located within 2 feet (610 mm) of a lot line are permitted to have roof eave projections not exceeding 4 inches (102 mm).
3. Foundation vents installed in compliance with this code are permitted.

Q: Do the requirements in Section R302.1 for projections apply to exterior balconies and decks as defined in Section R202?

A: Yes. Exterior balconies and decks that are attached to the building but not under roof are considered projections. Exterior balconies, decks and porches that are attached and under roof are considered to be a part of the primary building structure. Section 302.1 regulates projections into the fire separation distance. Projections can not extend more than 12 inches into this area. In addition, projections that do extend into an area where walls are required to be of one-hour fire-resistance construction must be protected on the underside with one-hour fire-resistance construction.

SECTION R308
GLAZING

R308.4 Hazardous locations. The following shall be considered specific hazardous locations for the purposes of glazing:

1. Glazing in swinging doors except jalousies.
2. Glazing in fixed and sliding panels of sliding door assemblies and panels in sliding and bifold closet door assemblies.
3. Glazing in storm doors.
4. Glazing in all unframed swinging doors.
5. Glazing in doors and enclosures for hot tubs, whirlpools, saunas, steam rooms, bathtubs and showers. Glazing in any part of a building wall enclosing these compartments where the bottom exposed edge of the glazing is less than 60 inches (1524 mm) measured vertically above any standing or walking surface.
6. Glazing, in an individual fixed or operable panel adjacent to a door where the nearest vertical edge is within a 24-inch (610 mm) arc of the door in a closed position and whose bottom edge is less than 60 inches (1524 mm) above the floor or walking surface.
7. Glazing in an individual fixed or operable panel, other than those locations described in Items 5 and 6 above, that meets all of the following conditions:
   - 7.1. Exposed area of an individual pane larger than 9 square feet (0.836 m^2).
   - 7.2. Bottom edge less than 18 inches (457 mm) above the floor.
   - 7.3. Top edge more than 36 inches (914 mm) above the floor.
   - 7.4. One or more walking surfaces within 36 inches (914 mm) horizontally of the glazing.
8. All glazing in railings regardless of an area or height above a walking surface. Included are structural baluster panels and nonstructural infill panels.
9. Glazing in walls and fences enclosing indoor and outdoor swimming pools, hot tubs and spas where the bottom edge of the glazing is less than 60 inches (1524 mm) above a walking surface and within 60 inches (1524 mm) horizontally of the water's edge. This shall apply to single glazing and all panes in multiple glazing.
10. Glazing adjacent to stairways, landings and ramps within 36 inches (914 mm) horizontally of a walking surface when the exposed surface of the glass is less than 60 inches (1524 mm) above the plane of the adjacent walking surface.
11. Glazing adjacent to stairways within 60 inches (1524 mm) horizontally of the bottom tread of a stairway in any direction when the exposed surface of the glass is less than 60 inches (1524 mm) above the nose of the tread.

Exception: The following products, materials and uses are exempt from the above hazardous locations:

1. Openings in doors through which a 3-inch (76 mm) sphere is unable to pass.
2. Decorative glass in Items 1, 6 or 7.
3. Glazing in Section R308.4, Item 6, when there is an intervening wall or other permanent barrier between the door and the glazing.
4. Glazing in Section R308.4, Item 6, in walls perpendicular to the plane of the door in a closed position, other than the wall toward which the door swings when opened, or where access through the door is to a closet or storage area 3 feet (914 mm) or less in depth. Glazing in these applications shall comply with Section R308.4, Item 7.
5. Glazing in Section R308.4, Items 7 and 10, when a protective bar is installed on the accessible side(s) of the glazing 36 inches ± 2 inches (914 mm ± 51 mm) above the floor. The bar shall be capable of withstanding a horizontal load of 50 pounds per linear foot (730 N/m) without contacting the glass and be a minimum of $1^1/_2$ inches (38 mm) in height.
6. Outboard panes in insulating glass units and other multiple glazed panels in Section R308.4, Item 7, when the bottom edge of the glass is 25 feet (7620 mm) or more above grade, a roof, walking surfaces, or other horizontal [within 45 degrees (0.79 rad) of horizontal] surface adjacent to the glass exterior.
7. Louvered windows and jalousies complying with the requirements of Section R308.2.
8. Mirrors and other glass panels mounted or hung on a surface that provides a continuous backing support.
9. Safety glazing in Section R308.4, Items 10 and 11, is not required where:
   - 9.1. The side of a stairway, landing or ramp has a guardrail or handrail, including balusters or in-fill panels, complying with the provisions of Sections 1013 and 1607.7 of the *International Building Code*; and
   - 9.2. The plane of the glass is more than 18 inches (457 mm) from the railing; or
   - 9.3. When a solid wall or panel extends from the plane of the adjacent walking surface to 34 inches (863 mm) to 36 inches (914 mm) above the floor and the construction at the top of that wall or panel is capable of withstanding the same horizontal load as the protective bar.
10. Glass block panels complying with Section R610.

Q: As defined in Section R308.4, Item 5 of the *International Residential Code*, is safety glazing required at glazed openings in building walls adjacent to the enclosure for hot tubs, whirlpools, saunas, steam rooms, bathtubs and showers where the glazing is not located within the area of the enclosure?

A: No. The provisions of Item 5 address only the requirements for safety glazing in the walls that create the enclosure surrounding the listed fixtures, and located within 60 inches vertically of the walking or standing surface within the fixture enclosure. Glazing located outside the area of the enclosing walls is not required to be safety glazing per the requirements of Item 5.

Q: Do the hazardous location provisions of Section R308.4, Item 6 apply to sliding door assemblies?

A: Yes. Item 6 applies to all doors. A sliding door assembly is a "door"; the 24-inch distance (arc) is measured from the vertical edge of both the fixed and sliding panels of sliding door assemblies.

SECTION R310
EMERGENCY ESCAPE AND RESCUE OPENINGS

R310.1 Emergency escape and rescue required. Basements and every sleeping room shall have at least one operable emergency and rescue opening. Such opening shall open directly into a public street, public alley, yard or court. Where basements contain one or more sleeping rooms, emergency egress and rescue openings shall be required in each sleeping room, but shall not be required in adjoining areas of the basement. Where emergency escape and rescue openings are provided they shall have a sill height of not more than 44 inches (1118 mm) above the floor. Where a door opening having a threshold below the adjacent ground elevation serves as an emergency escape and rescue opening and is provided with a bulkhead enclosure, the bulkhead enclosure shall comply with Section R310.3. The net clear opening dimensions required by this section shall be obtained by the normal operation of the emergency escape and rescue opening from the inside. Emergency escape and rescue openings with a finished sill height below the adjacent ground elevation shall be provided with a window well in accordance with Section R310.2. Emergency escape and rescue openings shall open directly into a public way, or to a yard or court that opens to a public way.

> **Exception:** Basements used only to house mechanical equipment and not exceeding total floor area of 200 square feet (18.58 m^2).

Q: In a second story sleeping room, are emergency escape and rescue windows permitted if removal of both the bottom and top sash is required to achieve the minimum net clear opening of 5.7 square feet?

A: No. The required opening size must be achieved through normal operation of the window without the removal of the top or bottom sash. Section R310.1 specifically states, "The net clear opening dimensions required by this section shall be obtained by the normal operation of the emergency escape and rescue opening from the inside." While most occupants are familiar with the normal operation by which to open the window, children and guests are frequently unfamiliar with special procedures necessary to remove the sashes. The time spent comprehending the special operation unnecessarily delays egress from the bedroom and could lead to panic and further confusion. Thus, windows that achieve the required opening dimensions only by having a special sequence of operations performed upon them, such as the removal of sashes or mullions, are not permitted.

SECTION R311
MEANS OF EGRESS

R311.4.3 Landings at doors. There shall be a floor or landing on each side of each exterior door. The floor or landing at the exterior door shall not be more than 1.5 inches (38 mm) lower than the top of the threshold. The landing shall be permitted to have a slope not to exceed 0.25 unit vertical in 12 units horizontal (2-percent).

Exceptions:

1. Where a stairway of two or fewer risers is located on the exterior side of a door, other than the required exit door, a landing is not required for the exterior side of the door provided the door, other than an exterior storm or screen door does not swing over the stairway.
2. The exterior landing at an exterior doorway shall not be more than $7^3/_4$ inches (196 mm) below the top of the threshold, provided the door, other than an exterior storm or screen door does not swing over the landing.
3. The height of floors at exterior doors other than the exit door required by Section R311.4.1 shall not be more than $7^3/_4$ inches (186 mm) lower than the top of the threshold.

The width of each landing shall not be less than the door served. Every landing shall have a minimum dimension of 36 inches (914 mm) measured in the direction of travel.

Q: Does Exception 2 in Section R311.4.3, which permits the landing to be $7^3/_4$ inches below the top of the threshold, apply to all exterior doors, including the required exit door?

A: Yes. The required exit door is classified as an exterior door. The provisions of Exception 2 in Section R311.4.3 permit a landing to be placed a maximum of $7^3/_4$ inches below the top of the threshold at any exterior doorway, which does include the required exit door.

SECTION R317
DWELLING UNIT SEPARATION

R317.1 Two-family dwellings. Dwelling units in two-family dwellings shall be separated from each other by wall and/or floor assemblies having not less than a 1-hour fire-resistance rating when tested in accordance with ASTM E 119. Fire-resistance-rated floor-ceiling and wall assemblies shall extend to and be tight against the exterior wall, and wall assemblies shall extend to the underside of the roof sheathing.

Exceptions:

1. A fire-resistance rating of $^1/_2$ hour shall be permitted in buildings equipped throughout with an automatic sprinkler system installed in accordance with NFPA 13.
2. Wall assemblies need not extend through attic spaces when the ceiling is protected by not less than $^5/_8$-inch (15.9 mm) Type X gypsum board and an attic draft stop constructed as specified in Section R502.12.1 is provided above and along the wall assembly separating the dwellings. The structural framing supporting the ceiling shall also be protected by not less than $^1/_2$-inch (12.7 mm) gypsum board or equivalent.

Q: Is a two-family dwelling with a property line separating the dwelling units required to comply with the separation provisions in Section R317.1?

A: No. The code does not address a property line within an attached two-family dwelling. A dwelling is a building that contains one or two dwelling units. A building line or property lines is a line established by law beyond which a building shall not extend. An attached two-family dwelling with a property line between the two dwelling units is considered two separate buildings, located on two separate lots. Two individual dwellings must comply with the fire separation distance required in Section R302.1

SECTION R318
MOISTURE VAPOR RETARDERS

R318.1 Moisture control. In all framed walls, floors and roof/ceilings comprising elements of the building thermal envelope, a vapor retarder shall be installed on the warm- in-winter side of the insulation.

Exceptions:

1. In construction where moisture or freezing will not damage the materials.
2. Where the framed cavity or space is ventilated to allow moisture to escape.
3. In counties identified as in climate zones 1 through 4 in Table N1101.2.

Q: The requirements in Section R318.1 stipulate that a vapor retarder must be installed on the warm-in-winter side of the thermal insulation. Is the installation of the vapor retarder required continuously across the face of the wall studs?

A: No. The word continuous is not used in the referenced code section. Therefore, a vapor retarder is not required to be installed in a manner that will provide a continuous barrier on the warm-in-winter side of the exterior wall. The performance criteria of a vapor retarder establishes that the material must have a permeance rating not exceeding 1 perm in order to limit the amount of moisture vapor that passes through the material.

CHAPTER 5

FLOORS

SECTION R502
WOOD FLOOR FRAMING

R502.7 Lateral restraint at supports. Joists shall be supported laterally at the ends by full-depth solid blocking not less than 2 inches (51 mm) nominal in thickness; or by attachment to a full-depth header, band or rim joist, or to an adjoining stud or shall be otherwise provided with lateral support to prevent rotation.

> **Exception:** In Seismic Design Categories D_0, D_1 and D_2, lateral restraint shall also be provided at each intermediate support.

Q: In Seismic Design Categories A, B and C, is solid blocking required where ends of floor joists are lapped at an intermediate bearing support?

A: No. In Seismic Design Categories A, B and C, solid blocking is not required between joists at an intermediate support where the joists are lapped or spliced as required in Section R502.6.1. The lap or splice at the intermediate support provides continuity of the joist, and therefore provides sufficient lateral support, as required in Section R502.7, to prevent rotation. Solid blocking, in addition to the lapped or spliced joists, is not required.

CHAPTER 6

WALL CONSTRUCTION

SECTION R602 WOOD WALL FRAMING

R602.3 Design and construction. Exterior walls of wood-frame construction shall be designed and constructed in accordance with the provisions of this chapter and Figures R602.3(1) and R602.3(2) or in accordance with AF&PA's NDS. Components of exterior walls shall be fastened in accordance with Table R602.3(1) through R602.3(4). Exterior walls covered with foam plastic sheathing shall be braced in accordance with Section R602.10. Structural sheathing shall be fastened directly to structural framing members.

REFERENCED SECTION:

R602.10.8 Connections. Braced wall panel sole plates shall be fastened to the floor framing and top plates shall be connected to the framing above in accordance with Table R602.3(1). Sills shall be fastened to the foundation or slab in accordance with Sections R403.1.6 and R602.11. Where joists are perpendicular to the braced wall lines above, blocking shall be provided under and in line with the braced wall panels.

Q: When a braced wall panel is connected to an elevated framed floor system as required in Section R602.10.8, the framed floor system is a structural member. Is the structural sheathing for a braced wall panel required to be attached to the floor framing system as required in Section R602.3 of the *International Residential Code* to complete the load path to the foundation?

A: No. The fastener requirements provided in the referenced tables are for the various structural framing members listed — and are prescriptive. Table R602.3(1) includes, but is not limited to, fasteners for the (wall) sole plate to the joist, wall studs to the sole plate, top plates to wall studs, 1-inch board sheathing to wall framing, and wood structural panel sheathing to wall framing. The load path is completed through the connection of the braced wall to the floor plate and the floor framing connection to the wall below and thus to the foundation. The prescriptive requirements do not require the structural sheathing extend beyond the bottom of the wall and be fastened directly to the floor framing at the perimeter rim joist.

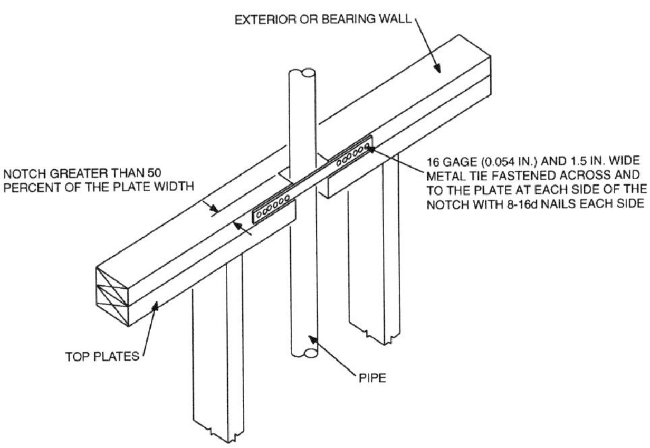

For SI: 1 inch = 25.4 mm.
NOTE: Condition for exterior and bearing walls.

FIGURE R602.6.1
TOP PLATE TO ACCOMMODATE PIPING

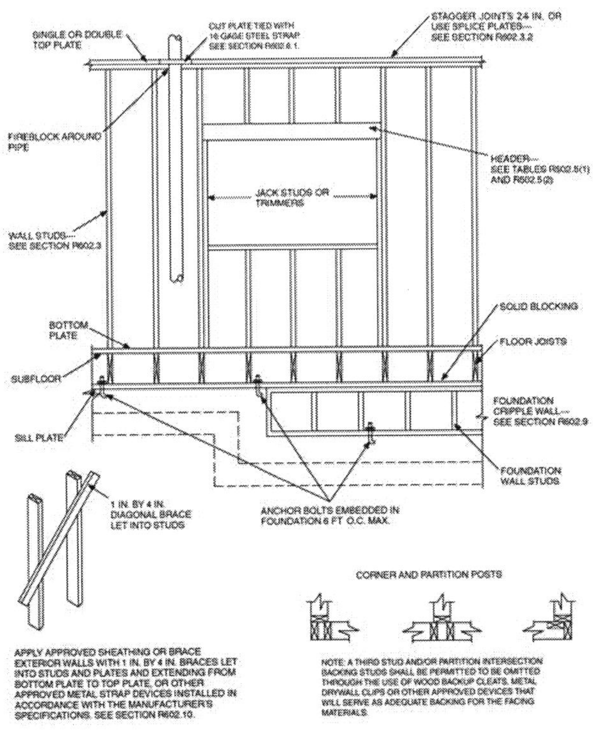

For SI: 1 inch = 25.4 mm, 1 foot = 304.8 mm.

FIGURE R602.3(2)
FRAMING DETAILS

The prescriptive requirements provided in the IRC do not address regions where the basic wind speeds equal or exceed 110 miles per hour. Structural framing systems in these high-wind regions must be engineered or designed to comply with one of the referenced standards listed in Section R301.2.1.1. The design for a specific framing system in a high-wind area may permit the sheathing to extend beyond the wall top plate and/or sole plate and fasten to the rim joist as an alternative to using metal straps, along with other methods, that may be necessary to tie the foundation, floor, wall, ceiling, and roof framing together.

R602.6.1 Drilling and notching of top plate. When piping or ductwork is placed in or partly in an exterior wall or interior load-bearing wall, necessitating cutting, drilling or notching of the top plate by more than 50 percent of its width, a galvanized metal tie of not less than 0.054 inch thick (1.37 mm) (16 ga) and 1 $^{1}/_{2}$ inches (38 mm) wide shall be fastened across and to the plate at each side of the opening with not less than eight 16d nails at each side or equivalent. See Figure R602.6.1.

Exception: When the entire side of the wall with the notch or cut is covered by wood structural panel sheathing.

Q: Where a wood stud wall is capped with a double top plate, is the 1.5-inch metal tie, as described, required on both of the top plates?

A: No. Where a double top plate is used, a single 1.5-inch metal tie is required on the uppermost plate only. No tie is required at the notched lower plate. This is illustrated in Figure R602.3(2) and Figure R602.6.1.

Q: Where a wood stud wall is capped with a double top plate, can a 3-inch wide metal tie with eight 16d nails on each side of the notch, equally distributed to both top plates, be used in lieu of the 1.5-inch metal tie prescribed?

A: Yes. The minimum required width of the metal tie is 1.5 inches. A metal tie that exceeds the required minimum width is not prohibited. It is not necessary to exceed the minimum number of required fasteners, because the width of the metal tie exceeds the required minimum.

R602.10 Wall bracing. All exterior walls shall be braced in accordance with this section. In addition, interior braced wall lines shall be provided in accordance with Section R602.10.1.1. For buildings in Seismic Design Categories D_0, D_1 and D_2, walls shall be constructed in accordance with the additional requirements of Sections R602.10.9, R602.10.11, and R602.11.

Q: Are all exterior wall lines required to be braced wall lines?

A: Yes. The *International Residential Code* is prescriptive. The provisions in Section R602.10 are prescriptive requirements which require all exterior walls to be braced in accordance with this section. All exterior walls must be constructed to include bracing that will resist the lateral loads resulting from wind or seismic forces, and to provide stability to the structure. The bracing must provide a complete path capable of transferring the loads from their point of origin through the load-resisting elements to the foundation.

Q: Is an interior braced wall line permitted to be braced with a single braced wall panel 48 inches in length?

A: Yes. Interior braced wall lines must comply with the same provisions as exterior braced wall lines. The amount and location of braced wall panels must comply with Table R602.10.1. The table requires, as a minimum, the wall bracing to be located at each end and at least every 25 feet on center. Section R602.10.1 modifies this by allowing the braced wall panel to "begin" up to 12.5 feet from the end of the braced wall line. Table R602.10.1 specifies the minimum amount of bracing to be provided as a percentage of the wall length based upon the basic wind speed, the seismic design category as well as the condition (location within the structure). If a single braced wall panel is not more than 12.5 feet from each end of the braced wall line and provides the required amount of bracing based on wind speed, seismic design category and condition, then it complies with this section of the code.

Q: Are interior braced wall lines required to comply with Section R602.10?

A: Yes. Interior braced wall lines must comply with the same provisions as exterior braced wall lines. The amount and location of braced wall panels must comply with Table R602.10.1.

R602.10.1 Braced wall lines. Braced wall lines shall consist of braced wall panel construction in accordance with Section R602.10.3. The amount and location of bracing shall be in accordance with Table R602.10.1 and the amount of bracing shall be the greater of that required by the seismic design category or the design wind speed. Braced wall panels shall begin no more than 12.5 feet (3810 mm) from each end of a braced wall line. Braced wall panels that are counted as part of a braced wall line shall be in line, except that offsets out-of-plane of up to 4 feet (1219 mm) shall be permitted provided that the total out-to-out offset dimension in any braced wall line is not more than 8 feet (2438 mm).

Q: Table R602.10.1 requires wall bracing to be located at each end and at least every 25 feet on center but not less than 16 percent of the braced wall line. When Methods 2, 3, 4, 6, 7 and 8 are used, each braced wall panel must be at least 48 inches in length. For a wall 25 feet in length, the 16 percent provision would permit the length of the braced wall panel to be a total of 4 feet. The braced wall panels are also permitted to begin no more than 12.5 feet from each end of braced wall line. Is a braced wall line permitted to be braced with a single braced wall panel 48 inches in length?

A: Yes. The amount and location of braced wall panels must comply with Table R602.10.1. The table requires, as a minimum, the wall bracing to be located at each end and at least every 25 feet on center. Section R602.10.1 modifies this by allowing the braced wall panel to "begin" up to 12.5 feet from the end of the braced wall line. Table R602.10.1 also specifies the minimum amount of bracing to be provided as a percentage of the wall length based upon the basic wind speed, the seismic design category as well as the condition (location within the structure). If a single braced wall panel is not more than 12.5 feet from each end of the braced wall line and provides the required amount of bracing based on wind speed, seismic design category and condition, then it complies with this section of the code.

Q: Does the section of wall that is 4 feet long and out-of-plane of the braced wall line contribute to the length of the braced wall line required by Table 602.10.1?

A: Yes. Out-of-plane braced wall panels that are offset no more 4 feet on either side of a braced wall line can only be counted as part of the braced wall line where the braced wall panel meets the length required by Section R602.10.4.

Q: Can an interior braced wall panel begin 12 feet away from an exterior braced wall line?

A: Yes. Regardless of whether the braced wall line is exterior or interior the provisions of Section R602.10.1 are applicable. Braced wall panels are not required to begin at the end of a braced wall line, they are permitted to begin up to 12.5 feet from each end of a braced wall line.

Q: Table R602.10.1 requires wall bracing to be located at each end and at least every 25 feet on center but not less than 16 percent of the braced wall line. When Methods 2, 3, 4, 6, 7 and 8 are used, each braced wall panel must be at least 48 inches in length. Is the total amount of wall bracing provided in a braced wall line permitted to be less than 8 feet in length?

A: Yes. The amount and location of braced wall panels must comply with Table R602.10.1. The table requires, as a minimum, the wall bracing to be located at each end and at least every 25 feet on center. Section R602.10.1 modifies this by allowing the braced wall panel to "begin" up to 12.5 feet from the end of the braced wall line. Table R602.10.1 specifies the minimum amount of bracing to be provided as a percentage of the wall length based upon the basic wind speed, the seismic design category as well as the condition (location within the structure). If a single braced wall panel is not more than 12.5 feet from each end of the braced wall line and provides the required amount of bracing based on wind speed, seismic design category and condition, then it complies with this section of the code.

R602.10.1.1 Spacing. Spacing of braced wall lines shall not exceed 35 feet (10 668 mm) on center in both the longitudinal and transverse directions in each story.

Exception: Spacing of braced wall lines not exceeding 50 feet shall be permitted where:

1. The wall bracing installed equals or exceeds the amount of bracing required by Table R602.10.1 multiplied by a factor equal to the braced wall line spacing divided by 35 feet and
2. The length-to-width ratio for the floor or roof diaphragm does not exceed 3:1.

Q: In order to increase the spacing of the braced wall lines from 35 feet to 50 feet, (1) the wall bracing must be increased by a factor equal to the spacing of the braced wall lines divided by 35 feet, and (2) the length-to-width ratio for the floor or roof diaphragm can not exceed 3:1. Is the factor and ratio determined using braced wall lines for the entire building?

A: No. If the spacing of the braced wall lines exceeds 50 feet, the provisions of Section R602.10.1.1 are not applicable. Should the spacing of the braced wall lines be between 35 feet and 50 feet, the factor and length-to-width ratio is based on the distance between the braced wall lines that bound the diaphragm, whether it is an interior braced wall line or an exterior braced wall line.

Q: Where exterior braced wall lines comply with the spacing requirements of Section R602.10.1.1, are interior braced wall lines required?

A: No. Interior braced wall lines are not required where all exterior braced wall lines are spaced a maximum of 35 feet on center as required by Section R602.10.1.1. However, the exception permits braced wall lines to be spaced a maximum of 50 feet on center if the amount of wall bracing is increased by a factor equal to the braced wall line spacing divided by 35 as required in Item 1, and the length-to-width ratio for the floor or roof diaphragm does not exceed 3:1 as required in Item 2.

R602.10.5 Continuous wood structural panel sheathing. When continuous wood structural panel sheathing is provided in accordance with Method 3 of Section R602.10.3 on all sheathable areas of all exterior walls, and interior braced wall lines, where required, including areas above and below openings, bracing wall panel lengths shall be in accordance with Table R602.10.5. Wood structural panel sheathing shall be installed at corners in accordance with Figure R602.10.5. The bracing amounts in Table R602.10.1 for Method 3 shall be permitted to be multiplied by a factor of 0.9 for wall with a maximum opening height that does not exceed 85 percent of the wall height or a factor of 0.8 for walls with a maximum opening height that does not exceed 67 percent of the wall height.

Q: When using continuous structural panel sheathing as prescribed in Section R602.10.5, are openings permitted in the wall at exterior corners?

A: No. The continuous structural panel sheathing requires wood structural panels to be installed at the corners.

SECTION R604
WOOD STRUCTURAL PANELS

R604.1 Identification and grade. Wood structural panels shall conform to DOC PS 1 or DOC PS 2. All panels shall be identified by a grade mark or certificate of inspection issued by an approved agency.

Q: Table R703.4 of the *International Residential Code* permits the use of plywood panels (exterior grade) as nonstructural weather-resistant exterior siding. Is a grade mark required on nonstructural plywood panel (exterior grade) siding?

A: Yes. The *International Residential Code* is a prescriptive code and does not differentiate between nonstructural and structural use for plywood in walls and partitions. All wood structural panels in walls and partitions must comply with Section R604 which states they shall conform to DOC PS 1 or DOC PS 2 and shall be identified by a grade mark or certificate issued by an approved agency.

CHAPTER 7

WALL COVERING

SECTION R703
EXTERIOR COVERING

R703.4 Attachments. Unless specified otherwise, all wall coverings shall be securely fastened in accordance with Table R703.4 or with other approved aluminum, stainless steel, zinc-coated or other approved corrosion-resistive fasteners. Where the basic wind speed per Figure R301.2(4) is 110 miles per hour (49 m/s) or higher, the attachment of wall coverings shall be designed to resist the component and cladding loads specified in Table R301.2(2), adjusted for height and exposure in accordance with Table R301.2(3).

Q: Table R703.4 of the *International Residential Code* permits the use of plywood panels (exterior grade) as nonstructural weather-resistant exterior siding. Is a grade mark required on nonstructural plywood panel (exterior grade) siding?

A: Yes. The *International Residential Code* is a prescriptive code and does not differentiate between nonstructural and structural use for plywood in walls and partitions. All wood structural panels in walls and partitions must comply with Section R604 which states they shall conform to DOC PS 1 or DOC PS 2 and shall be identified by a grade mark or certificate issued by an approved agency.

CHAPTER 8

ROOF-CEILING CONSTRUCTION

SECTION R801
GENERAL

R801.3 Roof Drainage. In areas where expansive or collapsible soils are known to exist, all dwellings shall have a controlled method of water disposal from roofs that will collect and discharge all roof drainage to the ground surface at least 5 feet (1524 mm) from foundation walls or to an approved drainage system.

Q: When a building utilizing a slab-on-grade, post-tensioned foundation is constructed on expansive soils, is water collected at the roof and discharged to the ground surface required to be discharged no closer than 5 feet from the edge of the slab-on-grade foundation?

A: No. Water collected at the roof and discharged at least 5 feet from foundation walls only. This provision does not apply to slab-on-grade foundations. Saturation of expansive soils by excessive water can lead to failure of the foundation because additional loads are imposed on the foundation wall by expansion of wet soil. Discharging roof water away from foundation walls serves to minimize saturation of expansive soils adjacent to the foundation walls.

CHAPTER 16

DUCT SYSTEMS

SECTION M1601 DUCT CONSTRUCTION

M1601.1.1 Above-ground duct systems. Above-ground duct systems shall conform to the following:

1. Equipment connected to duct systems shall be designed to limit discharge air temperature to a maximum of 250°F (121°C).
2. Factory-made air ducts shall be constructed of Class 0 or Class 1 materials as designated in Table M1601.1.1(1).
3. Fibrous duct construction shall conform to the SMACNA *Fibrous Glass Duct Construction Standards* or NAIMA *Fibrous Glass Duct Construction Standards*.
4. Minimum thickness of metal duct material shall be as listed in Table M1601.1.1(2). Galvanized steel shall conform to ASTM A 653.
5. Use of gypsum products to construct return air ducts or plenums is permitted, provided that the air temperature does not exceed 125°F (52°C) and exposed surfaces are not subject to condensation.
6. Duct systems shall be constructed of materials having a flame spread index not greater than 200.
7. Stud wall cavities and the spaces between solid floor joists to be used as air plenums shall comply with the following conditions:
   - 7.1. These cavities or spaces shall not be used as a plenum for supply air.
   - 7.2. These cavities or spaces shall not be part of a required fire-resistance-rated assembly.
   - 7.3. Stud wall cavities shall not convey air from more than one floor level.
   - 7.4. Stud wall cavities and joist-space plenums shall be isolated from adjacent concealed spaces by tight-fitting fire blocking in accordance with Section R602.8.

TABLE M1601.1.1(1)
CLASSIFICATION OF FACTORY-MADE AIR DUCTS

DUCT CLASS	MAXIMUM FLAME-SPREAD RATING
0	0
1	25

Q: Can plastic ducts be installed above-ground in residential applications?

A: Yes. Section M1601.1.1, Condition 2 permits the use of plastic air ducts for above-ground installation in residential applications; however, plastic air ducts must be manufactured from materials complying with Class 0 or Class 1 rating. In addition to the requirements of Condition 2, Condition 6 allows duct system components (other than the duct itself) to be constructed of materials having a flame-spread index not greater than 200. Plastic air ducts and duct system components that comply with these specific flame-spread requirements, as well as all other applicable provisions found elsewhere in the code, are permitted to be installed above-ground in residential applications.

INTERNATIONAL CODE INTERPRETATIONS

FIRE

2006

ICC
INTERNATIONAL CODE COUNCIL®

CHAPTER 1

ADMINISTRATION

SECTION 102
APPLICABILITY

102.4 Application of building code. The design and construction of new structures shall comply with the *International Building Code,* and any alterations, additions, changes in use or changes in structures required by this code, which are within the scope of the *International Building Code,* shall be made in accordance therewith.

Q: Is it the intent of the *International Fire Code* to permit the designated fire code official to apply the provisions of the *International Building Code* to existing buildings and structures (i.e., to make the *International Building Code* retroactive)?

A: No. It is not the intent of the *International Fire Code* to cause the *International Building Code* to be applied retroactively. The applicability of the *International Building Code* to existing structures is not determined within the context of the *International Fire Code* but rather is clearly established within the context of Chapter 34 of the *International Building Code*. Therefore, the *International Fire Code* is not structured to provide or determine the applicability of the *International Building Code* to existing buildings, but rather to provide for the proper functioning of fire protection systems, as well as reasonable housekeeping standards to prevent the likelihood of a fire. These areas are predominantly places where the *International Building Code* is not deemed applicable or retroactive. As a result, there are no conflicting lines of authority relative to existing buildings between the *International Building Code* and the *International Fire Code.*

SECTION 110
UNSAFE BUILDINGS

110.1 General. If during the inspection of a premises, a building or structure or any building system, in whole or in part, constitutes a clear and inimical threat to human life, safety or health, the code official shall issue such notice or orders to remove or remedy the conditions as shall be deemed necessary in accordance with this section and shall refer the building to the building department for any repairs, alterations, remodeling, removing or demolition required.

REFERENCED SECTION:

102.1 Construction and design provisions. The construction and design provisions of this code shall apply to:

1. Structures, facilities and conditions arising after the adoption of this code.
2. Existing structures, facilities and conditions not legally in existence at the time of adoption of this code.
3. Existing structures, facilities and conditions when identified in specific sections of this code.
4. Existing structures, facilities and conditions which, in the opinion of the fire code official, constitute a distinct hazard to life or property.

Q: Is Section 110.1 applicable to existing buildings?

A: Yes. According to Section 102.1, the code is applicable to existing structures, facilities and conditions. Section 110.1 refers to buildings or structures that are occupied. Once a certificate of occupancy has been issued, the building is considered an existing building.

CHAPTER 2

DEFINITIONS

SECTION 202
DEFINITIONS

OCCUPANCY CLASSIFICATION. For the purposes of this code, certain occupancies are defined as follows:

[B] Assembly Group A. Assembly Group A occupancy includes, among others, the use of a building or structure, or a portion thereof, for the gathering together of persons for purposes such as civic, social or religious functions; recreation, food or drink consumption; or awaiting transportation.

Exceptions:

1. A building used for assembly purposes with an occupant load of less than 50 persons shall be classified as a Group B occupancy.
2. A room or space used for assembly purposes with an occupant load of less than 50 persons and accessory to another occupancy shall be classified as a Group B occupancy or classified as part of that occupancy.
3. A room or space used for assembly purposes that is less than 750 square feet (70 m^2) in area and is accessory to another occupancy shall be classified as a Group B occupancy or classified as part of that occupancy.

Assembly occupancies shall include the following:

A-1 Assembly uses, usually with fixed seating, intended for the production and viewing of performing arts or motion pictures including but not limited to:

Motion picture theaters
Symphony and concert halls
Televison and radio studios admitting an audience
Theaters

A-2 Assembly uses intended for food and/or drink consumption including, but not limited to:

Banquet halls
Night clubs
Restaurants
Taverns and bars

A-3 Assembly uses intended for worship, recreation or amusement and other assembly uses not classified elsewhere in Group A, including, but not limited to:

Amusement arcades
Art galleries
Bowling alleys
Community halls
Courtrooms
Dance halls (not including food or drink consumption)
Exhibition halls
Funeral parlors
Gymnasiums (without spectator seating)
Indoor swimming pools (without spectator seating)
Indoor tennis courts (without spectator seating)
Lecture halls
Libraries
Museums
Places of religious worship
Pool and billiard parlors
Waiting areas in transportation terminals

A-4 Assembly uses intended for viewing of indoor sporting events and activities with spectator seating including, but not limited to:

Arenas
Skating rinks
Swimming pools
Tennis courts

A-5 Assembly uses intended for participation in or viewing outdoor activities including, but not limited to:

Amusement park structures
Bleachers
Grandstands
Stadiums

Q: Is a building or facility used for the showing, auctioning, and selling of nonhazardous merchandise classified as a Group A - Assembly Occupancy?

A: Yes. Any building or facility, or any room or space within a building or facility, used for the gathering of a prescribed number of people for amusement or recreation meets the definition for a Group A Occupancy. The use of a room or facility where people assemble for the purpose of conducting an auction is classified as a Group A-3 - Assembly Occupancy.

[B] Group I-2. This occupancy shall include buildings and structures used for medical, surgical, psychiatric, nursing or custodial care on a 24-hour basis of more than five persons who are not capable of self-preservation. This group shall include, but not be limited to, the following:

Hospitals
Nursing homes (both intermediate care facilities and skilled

nursing facilities)
Mental hospitals
Detoxification facilities

A facility such as the above with five or fewer persons shall be classified as Group R-3 or shall comply with the *International Residential Code* in accordance with Section 101.2 of the *International Building Code*.

A child care facility that provides care on a 24-hour basis to more than five children $2^1/_2$ years of age or less shall be classified as Group I-2.

Q: If the facility provides 24-hour accommodations to 10 persons, 5 of whom are not capable of self-preservation and 5 of whom are capable of responding to an emergency without assistance, is the facility classified as Group R-4?

A: No. A Group R-4 Residential Care/Assisted Living is a supervised residential facility that provides care for more than 5 occupants but not more than 16 occupants, all of whom must be "capable of responding to an emergency situation without physical assistance from staff." A facility that accommodates 5 or fewer such occupants is a Group R-3 occupancy, and a facility that accommodates more than 16 such occupants is a Group I-1 occupancy. A Group R-4 occupancy is not permitted to include any number of occupants that "are not capable of self-preservation."

A facility that accommodates more than 5 persons "who are not capable of self-preservation" is classified as a Group I-2 occupancy, while a facility that accommodates 5 or less occupants "who are not capable of self-preservation" is a Group R-3 occupancy.

A Group R-3 occupancy is permitted to accommodate a maximum of 5 occupants. The occupants may be "capable of responding to an emergency situation without physical assistance from staff" as permitted in a Group I-1 occupancy or "not capable of self-preservation" as required for a Group I-2 occupancy. A facility that accommodates 5 persons who are "capable of responding to an emergency situation without physical assistance from staff," **and** 5 persons "who are not capable of self-preservation" cannot be classified as a Group R-3 because the total occupant load of 10 persons exceeds the permitted maximum of 5 occupants.

The facility is a single occupancy; therefore, the entire facility must be assumed to be occupied by persons with the most restrictive capability when determining the occupancy classification of the facility. A facility that accommodates 10 occupants "who are not capable of self-preservation" is a Group I-2 occupancy.

Q: If an existing Group R-4 facility that provides 24-hour accommodations for 10 persons is expanded to accommodate an additional 6 persons who are not capable of responding to an emergency without physical assistance, must the occupancy classification of the existing facility be changed from a Group R-4 to Group I-2?

A: Yes. A facility that accommodates more than 5 occupants who are not capable of self preservation is classified as a Group I-2 occupancy. A Group R-4 facility is permitted to accommodate a maximum of 16 persons, but all of the occupants in a Group R-4 occupancy must be capable of self-preservation. The enlarged facility has an increased occupant load, which includes occupants "who are not capable of self-preservation"; therefore, the entire occupant load must be assumed to be persons with the most restrictive capability.

The existing Group R-4 facility may be expanded and retain the Group R-4 occupancy classification, but the expanded portion of the existing facility that accommodates the 6 additional occupants "who are not capable of self-preservation" must comply with all the technical provisions for a Group I-2 occupancy. Because the existing portion of the facility is a Group R-4 occupancy and the expanded portion of the facility is a Group I-2 occupancy the entire facility must comply with the provisions for mixed occupancies.

CHAPTER 5

FIRE SERVICE FEATURES

SECTION 503
FIRE APPARATUS ACCESS ROADS

503.2.3 Surface. Fire apparatus access roads shall be designed and maintained to support the imposed loads of fire apparatus and shall be surfaced so as to provide all-weather driving capabilities.

Q: The section says that fire apparatus roads must be "surfaced so as to provide all-weather driving capabilities." Does this require that the access roadway be paved?

A: No. Any stabilized access roadway that will carry the weight of the fire department apparatus is acceptable. The degree of effort required for the stabilization of the all-weather access roadway will depend on the size and weight of the fire department apparatus. The roadway must be stabilized to prevent the fire department apparatus from becoming stuck; and it must be maintained so that ruts, potholes, etc. do not prevent the access of the fire department apparatus.

CHAPTER 6

BUILDING SERVICES AND SYSTEMS

SECTION 605
ELECTRICAL EQUIPMENT, WIRING AND HAZARDS

605.7 Appliances. Electrical appliances and fixtures shall be tested and listed in published reports of inspected electrical equipment by an approved agency and installed and maintained in accordance with all instructions included as part of such listing.

REFERENCED SECTIONS:

SECTION 601
GENERAL

601.1 Scope. The provisions of this chapter shall apply to the installation, operation and maintenance of fuel-fired appliances and heating systems, emergency and standby power systems, electrical systems and equipment, mechanical refrigeration systems, elevator recall, stationary storage battery systems and commercial kitchen hoods.

INTERNATIONAL BUILDING CODE
SECTION 2701
GENERAL

[B] 2701.1 Scope. This chapter governs the electrical components, equipment and systems used in buildings and structures covered by this code. Electrical components, equipment and systems shall be designed and constructed in accordance with the provisions of the ICC *Electrical Code.*

Q: Is electrical equipment that is not part of a building's service equipment/system regulated by Section 605.7 of the *International Fire Code* and Section 2701.1 of the *International Building Code*?

A: No. Section 2701.1 of the *International Building Code* states that the provisions of the code shall control the design and construction of all new installations of electrical conductors, equipment and systems in buildings or structures. Section 605.7 of the *International Fire Code* requires electrical appliances and fixtures to be tested and listed. Only equipment and systems that are used as the building's electrical service and distribution system are intended to be controlled by the *International Codes.* All electrically controlled devices that are specifically designed and necessary for the general safety and welfare of the occupants, as well as other equipment and systems that are specifically addressed in the referenced standard, NFPA 70, would need to comply with the code provisions. The intended scope of Section 601.1 of the *International Fire Code* and Section 2701.1 of the *International Building Code* is the same as the intended scope of NFPA 70.

CHAPTER 9

FIRE PROTECTION SYSTEMS

SECTION 903
AUTOMATIC SPRINKLER SYSTEMS

903.2.1 Group A. An automatic sprinkler system shall be provided throughout buildings and portions thereof used as Group A occupancies as provided in this section. For Group A-1, A-2, A-3, and A-4 occupancies, the automatic sprinkler system shall be provided throughout the floor area where the Group A-1, A-2, A-3 or A-4 occupancy is located, and in all floors between the Group A occupancy and the level of exit discharge. For group A-5 occupancies, the automatic sprinkler system shall be provided in the spaces indicated in Section 903.2.1.5.

Q: Do the automatic sprinkler requirements in Section 903.2.1, based on the fire area of an occupancy, apply to buildings without surrounding exterior walls?

A: Yes. The entire area of a single floor of a building or structure must be considered when determining the automatic sprinkler requirements based on fire area. The entire area of a single floor of a building or structure may be considered a single fire area, or a portion of the floor area within a single story of a building or structure may be considered a fire area when separated in accordance with Section 706 of the *International Building Code*.

Where no surrounding exterior walls are provided along the perimeter of a building, the building area is used to identify and determine applicable fire areas. Section 502.1 of the *International Building Code* defines "Building Area" as: "The area included within surrounding exterior walls (or exterior walls and fire walls) exclusive of vent shafts and courts. Areas of the building not provided with surrounding walls shall be included in the building area if such areas are included within the horizontal projection of the roof or floor above."

Q: Do the automatic sprinkler requirements in Section 903.2.1, based on the fire area of an occupancy, apply to buildings without surrounding exterior walls if the building perimeter is enclosed with a barrier, such as fencing, rails or other obstruction that restricts exiting from any location along the perimeter of the building?

A: Yes. Means of egress is not a determining factor in identifying a fire area.

903.2.7 Group R. An automatic sprinkler system installed in accordance with Section 903.3 shall be provided throughout all buildings with a Group R fire area.

Q: Do the provisions of Section 903.2.7 in the *International Fire Code* require automatic sprinkler systems to be installed in all one- and two-family dwellings?

A: No. The provisions of the *International Fire Code* are applicable only to one- and two-family dwelling buildings constructed under the *International Building Code* and classified as Group R-3 by that code. They do not apply to detached one- and two-family dwellings and multiple single-family dwellings (townhouses) not more than three stories in height with a separate means of egress constructed under the *International Residential Code*.

The ICC Board of Directors, in originally approving the development of the *International Residential Code*, indicated that it is to be a stand-alone code for the construction of detached one- and two-family dwellings and multiple single-family dwellings (townhouses) not more than three stories in height with a seperate means of egress. The provisions for new construction that affect these type buildings are regulated exclusively by the *International Residential Code* and are not to be regulated by any other *International Code*. The exception to Section 101.2 of the *International Building Code*, stipulates that a detached one- and two-family dwellings and multiple single-family dwellings (townhouses) that is not more than three stories above grade plane in height shall comply with the *International Residential Code*. Likewise, Section R101.2 of the *International Residential Code* states that the provisions of the *International Residential Code* shall apply to the construction of detached one- and two-family dwellings and townhouses not more than three stories above grade plane in height with a separate means of egress and their accessory structures. Therefore, the construction of detached one- and two-family dwellings and multiple single-family dwellings (townhouses) within the scope of the *International Residential Code* is not required to comply with the provisions of either the *International Building Code* or the *International Fire Code*.

Where a dwelling or townhouse exceeds the allowable height in stories, does not provide individual egress for each dwelling unit, or does not conform to the prescriptive provisions of the *International Residential Code*, the structures are then beyond the scope of that code and its provisions cannot be applied. The building must then meet the provisions of the *International Building Code* for Group R-3 as well as the applicable provisions of the *International Fire Code*.

903.2.10.1 Stories and basements without openings. An automatic sprinkler system shall be installed in every story or basement of all buildings where the floor area exceeds 1,500 square feet (139.4 m^2) and where there is not provided at least one of the following types of exterior wall openings:

1. Openings below grade that lead directly to ground level by an exterior stairway complying with Section 1009 or an outside ramp complying with Section 1010. Openings shall be located in each 50 linear feet (15 240 mm), or fraction thereof, of exterior wall in the story on at least one side.
2. Openings entirely above the adjoining ground level totaling at least 20 square feet (1.86 m^2) in each 50 linear feet (15 240 mm), or fraction thereof, of exterior wall in the story on at least one side.

Q: An automotive service shop has below-grade service areas where employees perform oil changes and other minor maintenance services. The below-grade areas are open to the grade-level service bays, thus providing access to the underside of the vehicles without requiring the vehicle to be lifted into the air. Insofar as this below-grade space has no openings directly to the exterior, is it regulated as a windowless story and required to be equipped with an automatic fire suppression system in accordance with Section 903.2.10.1?

A: No. Due to the openness between the adjacent service levels, the below-grade area would be more appropriately regulated similar to a mezzanine rather than a story. A mezzanine is not regulated as a separate story but rather as part of the same story that it serves.

Therefore, if the below-grade service level is in compliance with the applicable provisions of Section 505, the windowless story provisions of Section 903.2.10.1 would be evaluated based on the exterior wall openings of the main level and not the service mezzanine below. The direct interconnections provided between two adjacent floor levels by multiple service openings provide access to the lower service area for fire-fighting and rescue operations. Due to the openness of the below-grade service level, it would not be regulated as a windowless story.

Q: A building is provided with openings in the exterior wall in lieu of the automatic fire suppression system in accordance with Section 903.2.10.1. Is the spacing between the jambs of adjacent openings in the exterior wall permitted to exceed 50 feet?

A: No. Section 903.2.10.1 requires that either exterior stairways, outside ramps or above-ground openings at least 20 square feet in size be located in each 50 lineal feet or fraction thereof of exterior walls. The required openings must be distributed such that the lineal distance between adjacent openings does not exceed 50 feet. The distribution of openings provides fire fighters with ready access to the interior of the building as well as multiple locations to vent smoke from the story in a fire situation.

If the openings in the exterior wall are located without regard to the location of adjacent openings, it is possible that segments of the exterior wall are not provided with the required access to the interior of the building for fire-fighting purposes. Any arrangement of required stairways, ramps or openings that results in a portion of the wall 50 feet or more in length with no openings to the exterior does not meet the intent of the code that access be provided in each 50 lineal feet.

903.3.1.2 NFPA 13R sprinkler systems. Where allowed in buildings of Group R, up to and including four stories in height, automatic sprinkler systems shall be installed throughout in accordance with NFPA 13R.

Q: Is an NFPA 13R sprinkler system permitted in a Group R occupancy where four stories of the building are located above grade plane and one story is located below grade plane?

A: Yes. The definition of "Building height" is the vertical distance from grade plane to the average height of the highest roof surface which establishes the grade plane as the reference point for determining building height and number of stories. Section 503 and Table 503 of the *International Building Code* establish the maximum building height and number of stories permitted for all occupancies above the grade plane. The maximum number of stories and building height in Table 503 of the *International Building Code* can be modified as permitted by Section 504.2 of the *International Building Code* when a sprinkler system is installed throughout in accordance with NFPA 13R. With the modification in Section 504.2, the building above the grade plane can be four stories, provided the overall building height does not exceed 60 feet measured to the average height of the highest roof surface. The type of occupancy permitted in the basement, including residential, is not controlled by the *International Fire Code* provided the basement is separated from the residential portions of the building in accordance with the mixed-use provisions of the *International Building Code*. Group S-2 occupancies can also be installed in the lower level of a Group R occupancy if they comply with the requirements of Section 508 of the *International Building Code*. The lower level, or basement, of the building must comply with the *International Building Code* definition for basement.

903.4.2 Alarms. Approved audible devices shall be connected to every automatic sprinkler system. Such sprinkler water-flow alarm devices shall be activated by water flow equivalent to the flow of a single sprinkler of the smallest orifice size installed in the system. Alarm devices shall be provided on the exterior of the building in an approved location. Where a fire alarm system is installed, actuation of the automatic sprinkler system shall actuate the building fire alarm system.

Q: Is the additional, supplemental interior sprinkler system alarm, which is required in Section 903.4.2, intended for general building evacuation?

A: No. The system alarm located inside the building is intended to alert occupants that water is flowing in the sprinkler system. The code does not require this to be a general evacuation alarm.

SECTION 905 STANDPIPE SYSTEMS

905.3.1 Building height. Class III standpipe systems shall be installed throughout buildings where the floor level of the highest story is located more than 30 feet (9144 mm) above the lowest level of the fire department vehicle access, or where the floor level of the lowest story is located more than 30 feet (9144 mm) below the highest level of fire department vehicle access.

Exceptions:

1. Class I standpipes are allowed in buildings equipped throughout with an automatic sprinkler system in accordance with Section 903.3.1.1 or 903.3.1.2.
2. Class I manual standpipes are allowed in open parking garages where the highest floor is located not more than 150 feet (45 720 mm) above the lowest level of fire department vehicle access.
3. Class I manual dry standpipes are allowed in open parking garages that are subject to freezing temperatures, provided that the hose connections are located as required for Class II standpipes in accordance with Section 905.5.
4. Class I standpipes are allowed in basements equipped throughout with an automatic sprinkler system.
5. In determining the lowest level of fire department vehicle access, it shall not be required to consider:
   - 5.1 Recessed loading docks for four vehicles or less, and
   - 5.2. Conditions where topography makes access from the fire department vehicle to the building impractical or impossible.

Q: Can a manual wet type water supply system be used in a Class I standpipe in a building that is not classified as a high-rise building?

A: Yes. Exception 1 to Section 905.3.1 permits the use of a Class I standpipe system in lieu of a Class III standpipe system when the building is equipped throughout with an automatic sprinkler system. However, Exception 1 does not stipulate a specific type of standpipe that must be used. The exception simply permits the use of a Class I standpipe. NFPA 14, as referenced in Section 905.2, provides specific design and installation requirements. A Class I standpipe system is intended for use by fire departments and trained personnel. A manual wet standpipe system requires the use of fire department equipment to pump water into the system to meet the demands put on the system.

SECTION 907 FIRE ALARM AND DETECTION SYSTEMS

907.2.4 Group F. A manual fire alarm system shall be installed in Group F occupancies that are two or more stories in height and have an occupant load of 500 or more above or below the lowest level of exit discharge.

Exception: Manual fire alarm boxes are not required where the building is equipped throughout with an automatic sprinkler system and the alarm notification appliances will activate upon sprinkler water flow.

Q: Given a building on grade, without a basement, see Attachment No. 1 Multi-Story Building. Is the second story of this building the first story above the lowest level of exit discharge?

A: No. When determining stories above the lowest level of exit discharge, a level, or floor level, is not a story. A level is the horizontal plane that is part of a story, not the entire story height. A story is the vertical space between the upper surface of one floor level and the upper surface of the floor level next above or below.

The level of exit discharge as shown in attached drawing of a Multi-Story Building is at elevation 0.0 feet the first floor level of the building is also at elevation 0.0 feet; therefore, the level of exit discharge and the first floor level of the building are at the same elevation. The first story of the building begins at elevation 0.0 feet (first floor) and extends to elevation 10.0 feet (second floor). The first story of the building is the first story above the level of exit discharge.

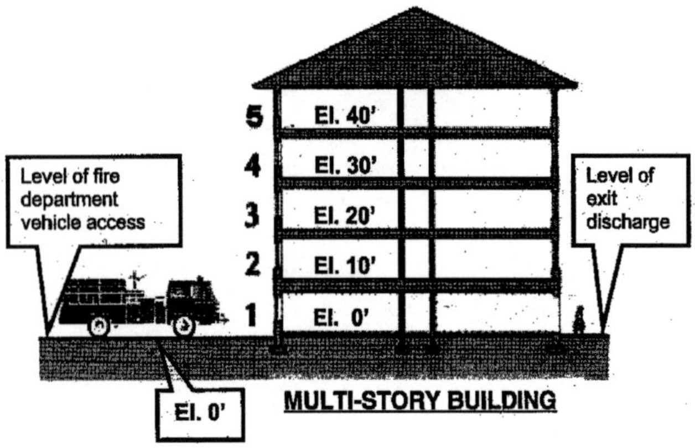

Attachment No. 1

Q: Is it the intent for the provisions of Section 907.2.4 of the *International Fire Code* to require a manual fire alarm system in a Group F occupancy where the occupant load of the first story of a multi-story building is 500 or more, and the first story is located at the lowest level of exit discharge?

A: No. The provisions of Section 907.2.4 do not require a manual fire alarm system in a Group F occupancy where the first story is the lowest level of exit discharge and the occupants have access to exit and exit discharge without being required to travel up or down an exit stairway. The provisions of Section 907.2.4 only apply to buildings two or more stories in height; they do not apply to a single-story building. A manual fire alarm system is required only when 500 or more occupants must travel up or down an exit stairway for no less than one story.

ICC Committee Interpretation 25-03, *International Fire Code* Section 907.2.4, issued May 11, 2004 addresses how to identify the location of stories relative to the level of exit discharge, and is not intended to address or clarify the specific scoping requirements for manual fire alarm systems in Group F occupancies.

This provision for a manual fire alarm system has its origin in the Standard Building Code, and used the term "above or below the street floor level," which was not defined. When the provision for a manual fire alarm system was brought forward into the working draft of the 2000 *International Building Code,* the term "above or below the level of exit discharge" was adopted; the original term "above or below the street floor level," used in the *Standard Building Code* was dropped. It was not the intent of the provisions in the *Standard Building Code* to require a manual fire alarm system when the minimum occupant load is located at or on the street floor level or on the "level of exit discharge."

In addition, "level of exit discharge" became a defined term in the *International Building Code* and no correlation was made to address the difference in the meaning and use of the two terms. As defined in the *International Building Code*, the "Level of exit discharge" is: "The horizontal plane located at the point at which an exit terminates and an exit discharge begins."

907.2.6 Group I. A manual fire alarm system shall be installed in Group I occupancies. An electrically supervised, automatic smoke detection system shall be provided in accordance with Sections 907.2.6.1 and 907.2.6.2.

> **Exception:** Manual fire alarm boxes in resident or patient sleeping areas of Group I-1 and I-2 occupancies shall not be required at exits if located at all nurses' control stations or other constantly attended staff locations, provided such stations are visible and continuously accessible and that travel distances required in Section 907.4.1 are not exceeded.

Q: Is an electrically supervised, smoke detection system required throughout a Group I-2 occupancy (hospital)?

A: No. Section 907.2.6 limits the need for electrically supervised, automatic smoke detection systems to waiting rooms open to corridors in all Group I occupancies, including Group I-2 hospitals. The provisions require only that a manual fire alarm system and an automatic fire detection system be installed in all occupancies classified as Group I.

907.2.6.2 Group I-2. Corridors in nursing homes (both intermediate care and skilled nursing facilities), detoxification facilities and spaces permitted to be open to the corridors by Section 407.2 of the *International Building Code* shall be equipped with an automatic fire detection system. Hospitals shall be equipped with smoke detection as required in Section 407.2 of the *International Building Code.*

Exceptions:

1. Corridor smoke detection is not required in smoke compartments that contain patient sleeping units where patient sleeping units are provided with smoke detectors that comply with UL 268. Such detectors shall provide a visual display on the corridor side of each patient sleeping unit and shall provide an audible and visual alarm at the nursing station attending each unit.
2. Corridor smoke detection is not required in smoke compartments that contain patient sleeping units where patient sleeping unit doors are equipped with automatic door-closing devices with integral smoke

detectors on the unit sides installed in accordance with their listing, provided that the integral detectors perform the required alerting function.

Q: Is an electrically supervised, smoke detection system required throughout a Group I-2 occupancy (hospital)?

A: No. Section 907.2.6.2 contains no requirements specific to a Group I-2 hospital. The provisions do require an automatic fire detection system in corridors and all spaces open to corridors in Group I-2 nursing homes and detoxification facilities. Section 907.2.6 requires installation of electrically supervised, automatic smoke detection systems only in waiting areas that are open to corridors in all Group I occupancies.

907.2.7 Group M. A manual fire alarm system shall be installed in Group M occupancies having an occupant load of 500 or more persons or more than 100 persons above or below the lowest level of exit discharge. The initiation of a signal from a manual fire alarm box shall initiate alarm notification appliances as required by Section 907.9.

Exceptions:

1. Covered mall buildings complying with Section 402.
2. Manual fire alarm boxes are not required where the building is equipped throughout with an automatic sprinkler system and the alarm notification appliances will automatically activate upon sprinkler water flow.

Q: Given a building on grade without a basement, see Attachment No. 1 Multi-Story Building. Is the second story of this building the first story above the lowest level of exit discharge?

A: No. When determining stories above the lowest level of exit discharge, a level, or floor level, is not a story. A level is the horizontal plane that is part of a story, not the entire story height. A story is the vertical space between the upper surface of one floor level and the upper surface of the floor level next above or below.

The level of exit discharge as shown in attached drawing of a Multi-Story Building is at elevation 0.0 feet, the first floor level of the building is also at elevation 0.0 feet; therefore, the level of exit discharge and the first floor level of the building are at the same elevation. The first story of the building begins at elevation 0.0 feet (first floor) and extends to elevation 10.0 feet (second floor). The first story of the building is the first story above the level of exit discharge.

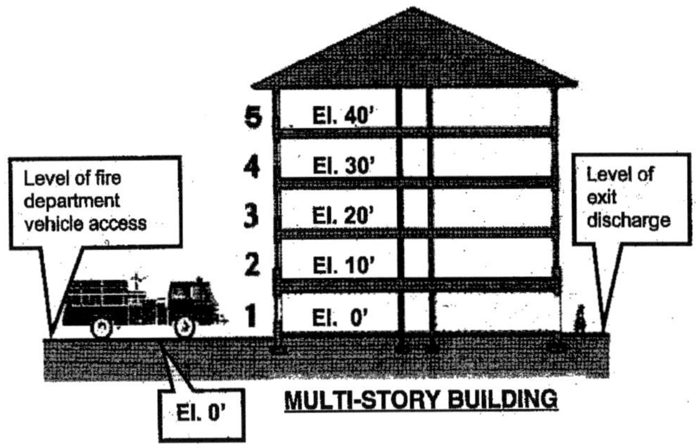

Attachment No. 1

Q: Is it the intent for the provisions of Section 907.2.7 of the *International Fire Code* to require a manual fire alarm system in a single-story Group M occupancy with an occupant load of less than 500 persons?

A: No. The provisions of Section 907.2.7 do not require a manual fire alarm system in a Group M occupancy where the first story is the lowest level of exit discharge and the occupants having access to exit and exit discharge without being required to travel up or down an exit stairway does not exceed 100 persons. A manual fire alarm system is required when a Group M occupancy has 500 or more occupants, regardless of where they are located within the building.

ICC Committee Interpretation 44-03, *International Fire Code*, Section 907.2.7, issued May 11, 2004 addresses how to identify the location of stories relative to the level of exit discharge, and is not intended to address or clarify the specific scoping requirements for manual fire alarm systems in Group M occupancies.

This provision for a manual fire alarm system has its origin in the *Standard Building Code*, and used the term "above or below the street floor level," which was not defined. When the provision for a manual fire alarm system was brought forward into the working draft of the 2000 *International Building Code* the term "above or below the level of exit discharge" was adopted; the original term "above or below the street floor level," used in the Standard Building Code was dropped. It was not the intent of the provisions in the *Standard Building Code* to require a manual fire alarm system when the minimum occupant load is located at or on the street floor level or on the "level of exit discharge."

In addition, "level of exit discharge" became a defined term in the *International Building Code* and no correlation was made to address the difference in the meaning and use of the two terms. As defined in the *International Building Code*, the "Level of exit discharge" is: "The horizontal plane located at the point at which an exit terminates and an exit discharge begins."

907.2.9 Group R-2. A manual fire alarm system shall be installed in Group R-2 occupancies where:

1. Any dwelling unit or sleeping unit is located three or more stories above the lowest level of exit discharge;
2. Any dwelling unit or sleeping unit is located more than one story below the highest level of exit discharge of exits serving the dwelling unit or sleeping unit; or
3. The building contains more than 16 dwelling units or sleeping units.

Exceptions:

1. A fire alarm system is not required in buildings not more than two stories in height where all dwelling units or sleeping units and contiguous attic and crawl spaces are separated from each other and public or common areas by at least 1-hour fire partitions and each dwelling unit or sleeping unit has an exit directly to a public way, exit court or yard.
2. Manual fire alarm boxes are not required throughout the building when the following conditions are met:
    - 2.1. The building is equipped throughout with an automatic sprinkler system in accordance with Section 903.3.1.1 or 903.3.1.2; and
    - 2.2. The notification appliances will activate upon sprinkler flow.
3. A fire alarm system is not required in buildings that do not have interior corridors serving dwelling units and are protected by an approved automatic sprinkler system installed in accordance with Section 903.3.1.1 or 903.3.1.2, provided that dwelling units either have a means of egress door opening directly to an exterior exit access that leads directly to the exits or are served by open-ended corridors designed in accordance with Section 1023.6, Exception 4.

Q: Given a building on grade without a basement, see Attachment No. 1 Multi-Story Building. Is the second story of this building the first story above the lowest level of exit discharge?

A: No. When determining stories above the lowest level of exit discharge, a level, or floor level, is not a story. A level is the horizontal plane that is part of a story, not the entire story height. A story is the vertical space between the upper surface of one floor level and the upper surface of the floor level next above or below.

The level of exit discharge as shown in attached drawing of a Multi-Story Building is at elevation 0.0 feet, the first floor level of the building is also at elevation 0.0 feet ; therefore, the level of exit discharge and the first floor level of the building are at the same elevation. The first story of the building begins at elevation 0.0 feet (first floor) and extends to elevation 10.0 feet (second floor). The first story of the building is the first story above the level of exit discharge.

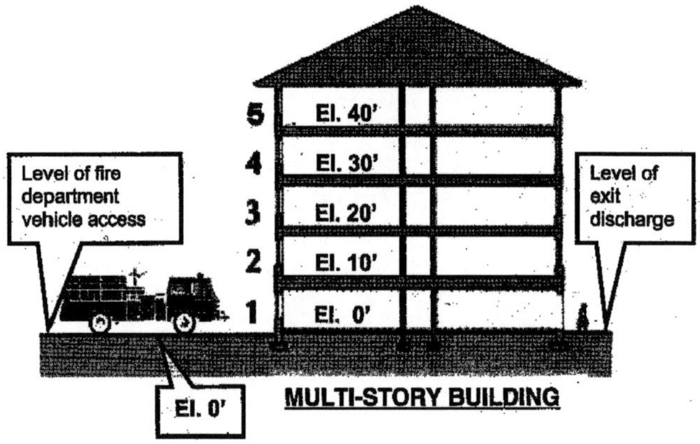

Attachment No. 1

907.2.10.1.2 Groups R-2, R-3, R-4 and I-1. Single- or multiple-station smoke alarms shall be installed and maintained in Groups R-2, R-3, R-4 and I-1 regardless of occupant load at all of the following locations:

1. On the ceiling or wall outside of each separate sleeping area in the immediate vicinity of bedrooms.
2. In each room used for sleeping purposes.
3. In each story within a dwelling unit, including basements but not including crawl spaces and uninhabitable attics. In dwellings or dwelling units with split levels and without an intervening door between the adjacent levels, a smoke alarm installed on the upper level shall suffice for the adjacent lower level provided that the lower level is less than one full story below the upper level.

Q: Do the provisions of Section 907.2.10.1.2 in the *International Fire Code* require smoke alarms to be installed in all new one- and two-family dwellings?

A: No. The smoke alarm provisions of the *International Fire Code* are only applicable to one- and two-family dwelling buildings constructed under the *International Building Code* and classified in Group R by that code. They do not apply to detached one- and two-family dwellings and townhouses not regulated as Group R occupancies and constructed under the *International Residential Code*.

The ICC Board of Directors, in originally approving the development of the *International Residential Code*, indicated that it is to be a stand-alone code for the construction of detached one- and two-family dwellings and multiple single-family dwellings (townhouses) not more than three stories in height with a seperate means of egress. The provisions for new construction that affect these type buildings are regulated exclusively by the *International Residential Code* and are not to be regulated by any other *International Code*. The Exception to Section 101.2 of the *International Building Code*, stipulates that a detached one- and two-family dwellings and multiple single-family dwellings (townhouses) that is not more than three stories above grade plane in height shall comply with the *International Residential Code*. Likewise, Section R101.2 of the *International Residential Code* states that the provisions of the *International Residential Code* shall apply to the construction of detached one- and two-family dwellings and multiple single-family dwellings (townhouses) not more than three stories above grade in height with a separate means of egress and their accessory structures. Therefore, the construction of detached one- and two-family dwellings and townhouses within the scope of the *International Residential Code* is not required to comply with the provisions of either the *International Building Code* or the *International Fire Code*.

Single- and multiple-station smoke alarm provisions for buildings not regulated in Occupancy Group R constructed under the *International Residential Code* are contained in Section R313 of that code and require that smoke alarms be installed in both new construction and buildings undergoing alterations, repairs or additions.

907.2.12.2 Emergency voice/alarm communication system. The operation of any automatic fire detector, sprinkler waterflow device or manual fire alarm box shall automatically sound an alert tone followed by voice instructions giving approved information and directions for a general or staged evacuation on a minimum of the alarming floor, the floor above and the floor below in accordance with the building's fire safety and evacuation plans required by Section 404. Speakers shall be provided throughout the building by paging zones. As a minimum, paging zones shall be provided as follows:

1. Elevator groups.
2. Exit stairways.
3. Each floor.
4. Areas of refuge as defined in Section 1002.1.

Exception: In Group I-1 and I-2 occupancies, the alarm shall sound in a constantly attended area and a general occupant notification shall be broadcast over the overhead page.

Q: Section 907.2.12.2 requires that the voice/alarm signaling system have the capability to sound automatically an alert signal to all occupants located within the building on a general or selective basis to specified terminal areas. Does the code require that the voice/alarm signaling system selectively identify and send messages to certain areas of the building?

A: No. The intent of these requirements is to provide the capability to send out selective messages to individual terminal areas of the building. This is not intended to prohibit the same message from going out to all terminal areas. However, the system must also contain controls within the fire command station to allow manual transmission of an evacuation signal and voice instructions on a selective *and* all-call basis to only the specified terminal areas.

Q: Section 907.2.12.2 identifies 4 different terminal areas requiring a voice/alarm signaling system to be established on a general or selective basis. Does this mean that there shall be a speaker in each of these areas to carry the voice communication for alarm systems?

A: Yes. The 4 areas identified are considered terminal areas. Each one is expected to have its own speakers, enabling selective messages to be provided.

907.3.2 Single- and multiple-station smoke alarms. Single- and multiple-station smoke alarms shall be installed in existing Group R occupancies in accordance with Sections 907.3.2.1 through 907.3.2.3.

Q: Do the provisions of Section 907.3.2 of the *International Fire Code* require smoke alarms to be installed in all existing one- and two-family dwellings?

A: No. The retroactive smoke alarm provisions of the *International Fire Code* are only applicable to existing one- and two-family dwelling buildings constructed under the *International Building Code* and classified in Group R-3 by that code. They do not apply to detached one- and two-family dwellings and multiple single-family dwellings (townhouses) not more than three stories in height with a separate means of egress constructed under and within the scope of the *International Residential Code*.

The ICC Board of Directors, in originally approving the development of the *International Residential Code*, indicated that it is to be a stand-alone code for the construction of detached one- and two-family dwellings and multiple single-family dwellings (townhouses) not more than three stories in height with a seperate means of egress. The provisions for new construction that affect these type buildings are regulated exclusively by the *International Residential Code* and are not to be regulated by any other *International Code*. The Exception to Section 101.2 of the *International Building Code*, stipulates that a detached one- and two-family dwellings and multiple single-family dwellings (townhouses) not more than three stories above grade plane in height shall comply with the *International Residential Code*. Likewise, Section R101.2 of the *International Residential Code* states that the provisions of the *International Residential Code* shall apply to the construction of detached one- and two-family dwellings and townhouses not more than three stories above grade in height with a separate means of egress and their accessory structures. Therefore, the construction of detached one- and two-family dwellings and townhouses within the scope of the *International Residential Code* is not required to comply with the provisions of either the *International Building Code* or the *International Fire Code*.

907.10.1 Visible alarms. Visible alarm notification appliances shall be provided in accordance with Sections 907.10.1.1 through 907.10.1.4.

Exceptions:

1. Visible alarm notification appliances are not required in alterations, except where an existing fire alarm system is upgraded or replaced, or a new fire alarm system is installed.
2. Visible alarm notification appliances shall not be required in exits as defined in Section 1002.1.

Q: Do the provisions of Sections 907.10.1 and 907.10.2 allow the option of providing either visible or audible alarm-notification appliances?

A: No. There is no option of providing either visible or audible alarms. If the building is a type that requires visible alarms, the devices must be heard as well. However, the code acknowledges the marginal value of audible alarms in certain areas where the average ambient noise is greater than 105 dBa. In these areas, only visible alarm-notification appliances in accordance with NFPA 72 are required.

907.10.2 Audible alarms. Audible alarm notification appliances shall be provided and sound a distinctive sound that is not to be used for any purpose other than that of a fire alarm. The audible alarm notification appliances shall provide a sound pressure level of 15 decibels (dBA) above the average ambient sound level or 5 dBA above the maximum sound level having a duration of at least 60 seconds, whichever is greater, in every occupied space within the building. The minimum sound pressure levels shall be: 70 dBA in occupancies in Groups R and I-1; 90 dBA in mechanical equipment rooms; and 60 dBA in other occupancies. The maximum sound pressure level for audible alarm notification appliances shall be 120 dBA at the minimum hearing distance from the audible appliance. Where the average ambient noise is greater than 105 dBA, visible alarm notification appliances shall be provided in accordance with NFPA 72 and audible alarm notification appliances shall not be required.

Exception: Visible alarm notification appliances shall be allowed in lieu of audible alarm notification appliances in critical care areas of Group I-2 occupancies.

Q: In an occupancy required to be equipped with a fire alarm system, is the audible alarm required to provide the minimum sound pressure level throughout bathrooms, mechanical equipment rooms and other such spaces of the building?

A: Yes. Section 907.10.2 requires that an audible alarm be heard in all occupied spaces of the building. The intent of the audible alarm is to warn occupants throughout the building of a fire condition. The alarm signal must be heard in rooms and spaces defined as "habitable" and "occupiable" as well as other spaces which are utilized by the occupants, such as bathrooms, storage rooms and mechanical rooms. Mechanical equipment rooms are specifically identified in this section due to the presence of equipment that, when operating, may necessitate an increase in the minimum sound pressure level based on higher ambient sound levels. Spaces that do not require audible warning include plenums, crawl spaces, attics and other spaces that are not intended, designed or arranged to be occupied or utilized by occupants. However, the code acknowledges the marginal value of audible alarms in certain areas where the average ambient noise is greater than 105 dBa. In these areas, only visible alarm notification appliances in accordance with NFPA 72 are required.

CHAPTER 10

MEANS OF EGRESS

[B] SECTION 1002 DEFINITIONS

[B] 1002.1 Definitions. The following words and terms shall, for the purposes of this chapter and as used elsewhere in this code, have the meanings shown herein.

FLOOR AREA, GROSS. The floor area within the inside perimeter of the exterior walls of the building under consideration, exclusive of vent shafts and courts, without deduction for corridors, stairways, closets, the thickness of interior walls, columns or other features. The floor area of a building, or portion thereof, not provided with surrounding exterior walls shall be the usable area under the horizontal projection of the roof or floor above. The gross floor area shall not include shafts with no openings or interior courts.

Q: Does "gross floor area" apply to all areas under the roof, including eaves and overhangs?

A: No. The gross floor area is the area within the inside perimeter of the exterior walls exclusive of areas open and unobstructed to the sky.

MEANS OF EGRESS. A continuous and unobstructed path of vertical and horizontal egress travel from any occupied portion of a building or structure to a public way. A means of egress consists of three separate and distinct parts: the exit access, the exit and the exit discharge.

Q: Are exterior stairways that are located anywhere between a building and the public way considered a part of the means of egress from a building?

A: Yes. The path of travel from the exit discharge doors to a public way is considered exit discharge in accordance with the definitions of "Means of egress" and "Exit discharge" in Section 1002. Any stairway constituting a portion of this path of travel would be considered an exit discharge stairway.

[B] SECTION 1003 GENERAL MEANS OF EGRESS

[B] 1003.1 Applicability. The general requirements specified in Sections 1003 through 1013 shall apply to all three elements of the means of egress system, in addition to those specific requirements for the exit access, the exit and the exit discharge detailed elsewhere in this chapter.

Q: Can the required capacity of each egress element from a room or space be determined by dividing the design occupant load of that space by the number of exits and exit access doorways that lead directly from the space?

A: Yes. Section 1003.1 requires the total capacity of all egress elements from a room or space to equal or exceed the design occupant load of the space as determined in accordance with Section 1004.1. Be aware that the design occupant load is the largest number of occupants derived by the application of Sections 1004.1.1 and also 1004.7.

An equal distribution of occupants between all egress elements leading out of a space reasonably approximates the use of multiple egress components from most occupancies should emergency evacuation be required. Note, however, that Section 1025.2 requires a more conservative egress distribution for assembly occupancies based on higher occupant load densities. Those elements that contribute to the required egress capacity from the space (e.g., exterior exit door, exit stairway or exit access doorway) must be identified with "Exit" signs as required by Section 1011.1 and comply with the other applicable provisions of Chapter 10.

Q: Is it mandatory for nonrequired doorways to contribute to the egress capacity of the room or space from which they lead?

A: No. The minimum required number of exits and exit access elements specified in Sections 1015.1 and 1019.1 provide an acceptable level of safety for the occupants. Although additional doorways are permitted, they are not specifically required to comply with all provisions of Chapter 10. Nonrequired doorways are to be constructured such that they serve their intended purpose without creating a hazard to the building occupants or general public. Accordingly, nonrequired doorways must not be identified as means of egress unless they fully comply with the applicable provisions of Chapter 10.

[B] 1003.4 Floor surface. Walking surfaces of the means of egress shall have a slip-resistant surface and be securely attached.

Q: A building contains exterior balconies, decks, porches and patios that are considered occupiable spaces as regulated by the code. Does Section 1003.4 require the walking surfaces of these building elements to be slip-resistant?

A: Yes. Because the pace of egress travel can become hurried during emergency situations, the probability of slipping on smooth or slick floor surfaces increases. To minimize the hazard, all floor surfaces in the means of egress are required to be slip resistant. As occupiable spaces, exterior balconies, decks, porches and patios are no different than interior building spaces, which are required to be provided with adequate access to the required exits. Accordingly, the occupants of the building's occupiable exterior elements must also have access to those same exits. The walking surfaces of those exterior elements are part of the means of egress, specifically exit access, and must have a slip-resistant surface. The code requires an acceptable minimum level of safety for all walking surfaces throughout the means of egress. The use of floor materials with highly polished, glazed, glossy or finely finished surfaces must be avoided.

[B] SECTION 1004 OCCUPANT LOAD

[B] 1004.1 Design occupant load. In determining means of egress requirements, the number of occupants for whom means of egress facilities shall be provided shall be determined in accordance with this section. Where occupants from accessory areas egress through a primary space, the calculated occupant load for the primary space shall include the total occupant load of the primary space plus the number of occupants egressing through it from the accessory area.

Q: The assembly area in a Group A-2 night club provides for seating at tables and chairs (unconcentrated), and also contains a dance floor. Is the area of the dance floor included in determining the minimum occupant load?

A: Yes. The code does not address the issue of non-simultaneous occupancy. Although the patrons on the dance floor could also have a seat at a table, the code requires the "net floor area" as defined by the code to be included in the calculation of the occupant load. Therefore, an occupant load must be determined for the seating area as well as the dance floor.

[B] 1004.1.1 Areas without fixed seating. The number of occupants shall be computed at the rate of one occupant per unit of area as prescribed in Table 1004.1.1. For areas without fixed seating, the occupant load shall not be less than that number determined by dividing the floor area under consideration by the occupant per unit of area factor assigned to the occupancy as set forth in Table 1004.1.1. Where an intended use is not listed in Table 1004.1.1, the building official shall establish a use based on a listed use that most nearly resembles the intended use.

> **Exception:** Where approved by the building official, the actual number of occupants for whom each occupied space, floor or building is designed, although less than those determined by calculation, shall be permitted to be used in the determination of the design occupant load.

Q: Can a floor area be designed for a higher occupant load than that determined by Table 1004.1.1?

A: Yes. Section 1004.1 establishes that the minimum design occupant load for which exits must be provided is the largest number computed from Sections 1004.1 through 1004.1.1. Section 1004.2 permits a building to be used and occupied by a higher occupant load, provided that the building fully complies with the code based on the higher number of occupants. (Note: The maximum occupant load, in any case, is not permitted to exceed one occupant per 7 square feet of occupiable floor space in accordance with Section 1004.2.)

[B] SECTION 1005 EGRESS WIDTH

[B] 1005.1 Minimum required egress width. The means of egress width shall not be less than required by this section. The total width of means of egress in inches (mm) shall not be less than the total occupant load served by the means of egress multiplied by the factors in Table 1005.1 and not less than specified elsewhere in this code. Multiple means of egress shall be sized such that the loss of any one means of egress shall not reduce the available capacity to less than 50 percent of the required capacity. The maximum capacity required from any story of a building shall be maintained to the termination of the means of egress.

> **Exception:** Means of egress complying with Section 1025.

Q: Is it the intent of Section 1005.1 to consider a pair of doors without a mullion as two separate openings for the purpose of determining the means of egress occupant capacity from Table 1005.1?

A: No. A pair of doors without a center mullion can be considered a single opening, provided that both doors are equipped with hardware complying with the requirements of Section 1008.1. The opening width to be used in computing the means of egress occupant capacity is the clear opening width with both doors open. (Note: One leaf must provide a minimum clear opening width of 32 inches in accordance with Section 1008.1.1 for those occupancies requiring a door opening to have a 32-inch clear width.)

[B] SECTION 1007 ACCESSIBLE MEANS OF EGRESS

[B] 1007.2.1 Elevators required. In buildings where a required accessible floor is four or more stories above or below a level of exit discharge, at least one required accessible means of egress shall be an elevator complying with Section 1007.4.

Exceptions:

1. In buildings equipped throughout with an automatic sprinkler system installed in accordance with Section 903.3.1.1 or 903.3.1.2, the elevator shall not be required on floors provided with a horizontal exit and located at or above the level of exit discharge.
2. In buildings equipped throughout with an automatic sprinkler system installed in accordance with Section 903.3.1.1 or 903.3.1.2, the elevator shall not be required on floors provided with a ramp conforming to the provisions of Section 1010.

Q: Given a building on grade, without a basement, see Attachment No. 1 Multi-Story Building. Is the second story of this building the first story above the lowest level of exit discharge?

A: No. When determining stories above the lowest level of exit discharge, a level, or floor level, is not a story. A "level" is the horizontal plane that is part of a story, not the entire story height. A "story" is the vertical space between the upper surface of one floor level and the upper surface of the floor level next above or below.

The level of exit discharge as shown in attached drawing of a Multi-Story Building is at elevation 0.0 feet, the first floor level of the building is also at elevation 0.0 feet; therefore, the level of exit discharge and the first floor level of the building are at the same elevation. The first story of the building begins at elevation 0.0 feet (first floor) and extends to elevation 10.0 feet (second floor). The first story of the building is the first story above the level of exit discharge.

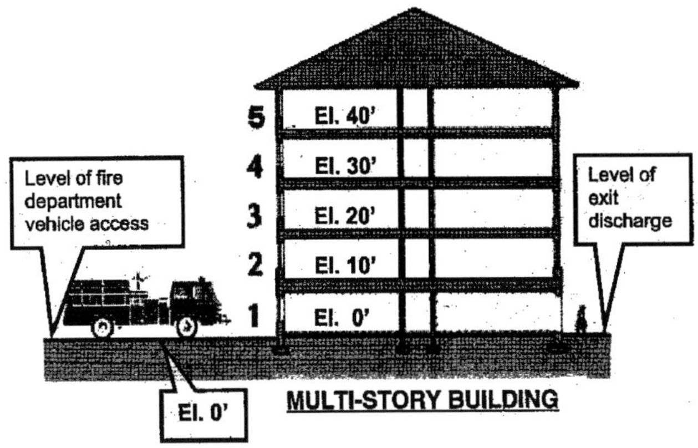

Attachment No. 1

[B] SECTION 1008 DOORS, GATES AND TURNSTILES

[B] 1008.1.2 Door swing. Egress doors shall be side-hinged swinging.

Exceptions:

1. Private garages, office areas, factory and storage areas with an occupant load of 10 or less.
2. Group I-3 occupancies used as a place of detention.
3. Critical or intensive care patient rooms within suites of health care facilities.
4. Doors within or serving a single dwelling unit in Groups R-2 and R-3.
5. In other than Group H occupancies, revolving doors complying with Section 1008.1.3.1.
6. In other than Group H occupancies, horizontal sliding doors complying with Section 1008.1.3.3 are permitted in a means of egress.
7. Power-operated doors in accordance with Section 1008.1.3.2.
8. Doors serving a bathroom within an individual sleeping unit in Group R-1.

Doors shall swing in the direction of egress travel where serving an occupant load of 50 or more persons or a Group H occupancy.

The opening force for interior side-swinging doors without closers shall not exceed a 5-pound (22 N) force. For other side-swinging, sliding and folding doors, the door latch shall release when subjected to a 15-pound (67 N) force. The door shall be set in motion when subjected to a 30-pound (133 N) force. The door shall swing to a full-open position when subjected to a 15-pound (67 N) force. Forces shall be applied to the latch side.

Q: A room with an occupancy in Group A, B, E, F or M with an occupant load of more than 49 individuals is required by Table 1015.1 to have two means of egress doors. Are these doors required to swing in the direction of egress travel?

A: Yes. Section 1008.1.2 requires all doors serving an occupant load of 50 or more or a high-hazard occupancy to swing in the direction of egress travel. The occupant load of 50 or more relates to the occupant load of the room and not to the exiting capacity of the doors.

[B] 1008.1.3.2 Power-operated doors. Where means of egress doors are operated by power, such as doors with a photoelectric-actuated mechanism to open the door upon the approach of a person, or doors with power-assisted manual operation, the design shall be such that in the event of power failure, the door is capable of being opened manually to permit means of egress travel or closed where necessary to safeguard means of egress. The forces required to open these doors manually shall not exceed those specified in Section 1008.1.2, except that the force to set the door in motion shall not exceed 50 pounds (220 N). The door shall be capable of swinging from any position to the full width of the opening in which such door is installed when a force is applied to the door on the side from which egress is made. Full-power-operated doors shall comply with BHMA A156.10. Power-assisted and low-energy doors shall comply with BHMA A156.19.

Exceptions:

1. Occupancies in Group I-3.
2. Horizontal sliding doors complying with Section 1008.1.3.3.
3. For a biparting door in the emergency breakout mode, a door leaf located within a multiple-leaf opening shall be exempt from the minimum 32-inch (813 mm) single-leaf requirement of Section 1008.1.1, provided a minimum 32-inch (813 mm) clear opening is provided when the two biparting leaves meeting in the center are broken out.

Q: Can the opening that results from the emergency swing of adjacent panels of power-operated sliding or swinging doors be used to meet the minimum required clear width?

A: Yes. Section 1008.1.3.2 deals with the maximum force to be applied to power-operated doors that will permit them to be considered as an egress element. These forces must open the normally power-operated panels in a swing mode during a loss of power.

Where Section 1008.1.1 requires a minimum clear width of 32 inches and a door opening includes two door leaves without a mullion, one leaf shall provide the clear opening width of 32 inches, regardless of the door operation.

[B] 1008.1.3.3 Horizontal sliding doors. In other than Group H occupancies, horizontal sliding doors permitted to be a component of a means of egress in accordance with Exception 5 to Section 1008.1.2 shall comply with all of the following criteria:

1. The doors shall be power operated and shall be capable of being operated manually in the event of power failure.
2. The doors shall be openable by a simple method from both sides without special knowledge or effort.
3. The force required to operate the door shall not exceed 30 pounds (133 N) to set the door in motion and 15 pounds (67 N) to close the door or open it to the minimum required width.
4. The door shall be openable with a force not to exceed 15 pounds (67 N) when a force of 250 pounds (1100 N) is applied perpendicular to the door adjacent to the operating device.
5. The door assembly shall comply with the applicable fire protection rating and, where rated, shall be self-closing or automatic-closing by smoke detection, shall be installed in accordance with NFPA 80 and shall comply with Section 715.
6. The door assembly shall have an integrated standby power supply.
7. The door assembly power supply shall be electrically supervised.
8. The door shall open to the minimum required width within 10 seconds after activation of the operating device.

Q: Does Section 1008.3 apply to security gates or grilles installed on the inside of means of egress doorways and that swing against the direction of egress travel if the gates are left open during occupancy?

A: No. Such a security gate can be dealt with under Section 1008.1.3.5.

Q: If gates are permitted, can they be equipped with either double-cylinder deadbolts requiring a key from both sides, or padlock and hasp?

A: No. A double-cylinder deadbolt or other type of key locking device is prohibited on egress doors in accordance with Section 1008.1.8 and would, therefore, not be permissible on a security gate in a doorway.

Q: Does Section 1008.1.3.3 permit the use of nonpower-operated horizontal sliding doors as a means of egress if the door, when pushed, will swing in the event of an emergency?

A: No. Except as provided for in the exceptions, Section 1008.1.3.3 requires all nonpower-operated egress doors to be of a side-swinging type. Furthermore, the door must swing in the direction of egress whenever serving an occupant load of 50 or more or a high-hazard occupancy. The normal operation of a horizontal sliding door is not of a side-swinging type. Although the door can swing open in the event of an emergency, the operation will likely not be known to the building occupants because the normal operation is to open horizontally.

Criteria for Section 1008.1.3.3 permit the use of horizontal sliding doors that comply with the conditions set forth therein. Power-operated sliding or swinging doors are required to comply with Section 1008.1.3.2 provisions.

[B] 1008.1.3.4 Access-controlled egress doors. The entrance doors in a means of egress in buildings with an occupancy in Group A, B, E, M, R-1 or R-2 and entrance doors to tenant spaces in occupancies in Groups A, B, E, M, R-1 and R-2 are permitted to be equipped with an approved entrance and egress access control system which shall be installed in accordance with all of the following criteria:

1. A sensor shall be provided on the egress side arranged to detect an occupant approaching the doors. The doors shall be arranged to unlock by a signal from or loss of power to the sensor.
2. Loss of power to that part of the access control system which locks the doors shall automatically unlock the doors.
3. The doors shall be arranged to unlock from a manual unlocking device located 40 inches to 48 inches (1016 mm to 1219 mm) vertically above the floor and within 5 feet (1524 mm) of the secured doors. Ready access shall be provided to the manual unlocking device and the device shall be clearly identified by a sign that reads "PUSH TO EXIT." When operated, the manual unlocking device shall result in direct interruption of power to the lock—independent of the access control system electronics—and the doors shall remain unlocked for a minimum of 30 seconds.
4. Activation of the building fire alarm system, if provided, shall automatically unlock the doors, and the doors shall remain unlocked until the fire alarm system has been reset.
5. Activation of the building automatic sprinkler or fire detection system, if provided, shall automatically unlock the doors. The doors shall remain unlocked until the fire alarm system has been reset.
6. Entrance doors in buildings with an occupancy in Group A, B, E or M shall not be secured from the egress side during periods that the building is open to the general public.

Q: Groups A, B, E and M occupancies, if criteria 1 through 5 are met and the access-controlled egress doors unlock on the egress side prior to an occupant reaching the doors and no subsequent securing devices are installed, does the access-control system need to be disabled during the hours the building is open to the public?

A: No, the access-controlled system at egress doors is permitted to remain operational at all times; however, Item 6 specifically requires that during the hours the building is open to the general public, egress doors equipped with a controlled-access system shall not be secured from the egress side. It is permissible to lock occupants out of the building, but occupants may not be locked in the building. This restriction applies to Group A, B, E and M occupancies.

Q: Are the requirements for access-controlled egress doors described in Section 1008.1.3.4 intended as an alternate method to panic hardware equipped with actuating switches?

A: No. The access-controlled egress requirements in Section 1008.1.3.4 are not permitted to serve as an alternative method for egress hardware required elsewhere in the code. The requirements in Section 1008.1.3.4 provide additional controls beyond the minimum requirements, which may be added to control access through an entrance door that also serves as an egress or exit door. Where panic hardware is installed and secured during periods when the building is open to the public, the panic hardware itself and the electronic actuating switches within the panic hardware must be operational for egress as required in Items 1 through 6.

[B] 1008.1.8 Door operations. Except as specifically permitted by this section egress doors shall be readily openable from the egress side without the use of a key or special knowledge or effort.

Q: Section 1008.1.8 states that all means of egress doors must be readily openable from the side from which egress is to be made without the use of a key, special knowledge or effort. Where there is a series of adjacent doors for convenience but only one of them is required to satisfy the means of egress width based on the occupant load, can all but one be provided with a lock requiring the use of a key? The remaining door is equipped with a latch, that is readily recognized as being latched, and its operation does not require special knowledge or effort.

A: Yes. If supplemental adjacent doors are provided in excess of those required to satisfy the means of egress width based on the occupant load, the additional doors are not required to comply with the requirements for egress doors, provided that a means of channeling the natural flow of traffic to the latched door through the use of railings, ropes or other equivalent means is employed. The required egress door must also be clearly marked in accordance with Section 1011.

[B] 1008.1.8.3 Locks and latches. Locks and latches shall be permitted to prevent operation of doors where any of the following exists:

1. Places of detention or restraint.
2. In buildings in occupancy Group A having an occupant load of 300 or less, Groups B, F, M and S, and in places of religious worship, the main exterior door or doors are permitted to be equipped with key-operated locking devices from the egress side provided:
   - 2.1. The locking device is readily distinguishable as locked,
   - 2.2. A readily visible durable sign is posted on the egress side on or adjacent to the door stating: THIS DOOR TO REMAIN UNLOCKED WHEN BUILDING IS OCCUPIED. The sign shall be in letters 1 inch (25 mm) high on a contrasting background,
   - 2.3. The use of the key-operated locking device is revokable by the building official for due cause.
3. Where egress doors are used in pairs, approved automatic flush bolts shall be permitted to be used, provided that the door leaf having the automatic flush bolts has no doorknob or surface-mounted hardware.
4. Doors from individual dwelling or sleeping units of Group R occupancies having an occupant load of 10 or less are permitted to be equipped with a night latch, dead bolt or security chain, provided such devices are openable from the inside without the use of a key or tool.

Q: The only means of access to and egress from an exterior balcony that serves a dwelling unit requires an occupant to travel through the dwelling unit. Is the door separating the balcony from the interior of the dwelling unit permitted to be locked from the interior?

A: Yes. Exception 4 to Section 1008.1.8.3 allows egress doors from an individual dwelling unit with an occupant load of 10 or less to be equipped with a night latch, dead bolt or security chain to maintain the security of the occupants. Similarly, the egress door from the balcony is permitted to be equipped with any of the types of security measures identified in Exception 4 or hardware that is locked manually from the interior. Self-locking hardware that could accidentally and automatically trap an occupant on the balcony is prohibited.

Q: Are the provisions of Section 1008.1.8.3, Exception 2, of the *International Building Code* applicable to buildings with one exit as regulated in Section 1014.1?

A: Yes. The requirements of Section 1008.1.8.3, Exception 2, establish specific conditions under which a building's "main exterior" means of egress door may be equipped with locking hardware that requires the use of a key to operate but do not limit their application to any particular building configuration. Because the definition of "Means of egress door" does not include the descriptor "main exterior," it must assume the ordinarily accepted meaning (see Section 201.4) which, in this case, would be the principal means of egress door located in the exterior wall of a building and would likely also be considered the principal means of entrance. Because a building with one exit as regulated by Section 1015.1 has but a single exterior means of egress door, it must naturally be considered the "main" door to which the provisions of Section 1008.1.8.3, Exception 2, are applicable.

Q: The main exterior doors of a Group A occupancy with an occupant load of 300 or less is permitted by the provisions of the *International Building Code* Section 1008.1.8.3, Exception 2, to be equipped with a key-operated locking device that, in the unlocked position, will allow the door to be readily openable.

Is the main exterior egress door in a Group A occupancy with an occupant load of 100 or more required to be equipped with panic hardware or fire exit hardware, provisioned in Section 1008.1.9, in addition to the key-operated locking device permitted in Section 1008.1.8.3?

A: No. Panic hardware or fire exit hardware as described in Section 1008.1.9 is not required on the main exterior egress door of a Group A occupancy where the door is equipped with a single key-operated locking device that complies with all the provisions of Section 1008.1.8.3, Exception 2. A single key-operated locking device is permitted as an exception or alternative to the primary requirements for locking or latching of egress doors. The permissive requirements of Section 1008.1.8.3, Exception 2 are very specific in identifying when and under what conditions the exception may be applied. The main exterior door that is permitted to be equipped with a keyed lock is also required to be unlocked while the building is occupied. A single locking device in the unlocked position will allow the door to swing freely, and panic hardware or fire exit hardware is only required on a door that is held in the closed position by a latch or lock.

Q: With the exception of doors requiring panic hardware in accordance with Section 1008.1.9, would all required egress doors be subject to the provisions of Section 1008.1.8.3, independent of the number of occupants occupying the building or space?

A: Yes. The provisions of Section 1008.1.8.3 make no exceptions for the number of occupants involved. However, Section 1008.1.8.3 provides for certain exceptions. The provisions of Section 1008.1.9 for panic hardware are applicable as indicated for the occupancies and occupant load stipulated.

[B] 1008.1.8.6 Delayed egress locks. Approved, listed, delayed egress locks shall be permitted to be installed on doors serving any occupancy except Group A, E and H occupancies in buildings that are equipped throughout with an automatic sprinkler system in accordance with Section 903.3.1.1 or an approved automatic smoke or heat detection system installed in accordance with Section 907, provided that the doors unlock in accordance with Items 1 through 6 below. A building occupant shall not be required to pass through more than one door equipped with a delayed egress lock before entering an exit.

1. The doors unlock upon actuation of the automatic sprinkler system or automatic fire detection system.
2. The doors unlock upon loss of power controlling the lock or lock mechanism.
3. The door locks shall have the capability of being unlocked by a signal from the fire command center.
4. The initiation of an irreversible process which will release the latch in not more than 15 seconds when a force of not more than 15 pounds (67 N) is applied for 1 second to the release device. Initiation of the irreversible process shall activate an audible signal in the vicinity of the door. Once the door lock has been released by the application of force to the releasing device, relocking shall be by manual means only.

   Exception: Where approved, a delay of not more than 30 seconds is permitted.
5. A sign shall be provided on the door located above and within 12 inches (305 mm) of the release device reading: PUSH UNTIL ALARM SOUNDS. DOOR CAN BE OPENED IN 15 [30] SECONDS.
6. Emergency lighting shall be provided at the door.

Q: Do the requirements in Section 1008.1.8.6, Item 4 require both labeled fire doors and nonlabeled doors to be equipped with latching devices?

A: Yes. All the provisions in Section 1008.1.8.6 shall apply to any door where a delayed egress lock is used. The "latch" referenced in Item 4 is a requirement of the delayed egress locking mechanism (see Item 2), and is required to be installed as part of the locking system on both labeled fire door assemblies and nonrated doors.

A delayed egress locking system, as defined in Section 1008.1.8.6, is a nonmandatory locking system that is permitted to be used on egress doors; however, it must be used in conjunction with and in addition to locks and latches required elsewhere in the code. The delayed egress lock provides an approved means of restraining/releasing an egress door that is otherwise required to remain operable from the egress side.

[B] 1008.1.9 Panic and fire exit hardware. Where panic and fire exit hardware is installed, it shall comply with the following:

1. The actuating portion of the releasing device shall extend at least one-half of the door leaf width.
2. The maximum unlatching force shall not exceed 15 pounds (67 N).

Each door in a means of egress from a Group A or E occupancy having an occupant load of 50 or more and any Group H occupancy shall not be provided with a latch or lock unless it is panic hardware or fire exit hardware.

Exception: A main exit of a Group A occupancy in compliance with Section 1008.1.8.3, Item 2.

Electrical rooms with equipment rated 1,200 amperes or more and over 6 feet (1829 mm) wide that contain overcurrent devices, switching devices or control devices with exit access doors must be equipped with panic hardware and doors must swing in the direction of egress.

If balanced doors are used and panic hardware is required, the panic hardware shall be the push-pad type and the pad shall not extend more then one-half the width of the door measured from the latch side.

Q: Can a thumb-turn cylinder lock or head and floor bolts be installed on a means of egress door required to be equipped with panic hardware in accordance with Section 1008.1.9?

A: No. The function of panic hardware is to make the door free swinging upon its operation. The proposed hardware would eliminate the opportunity for emergency egress in a panic situation. Note: Section 1008.1.8.4 prohibits manually operated flush or surface bolts.

Q: Does Section 1008.1.9 require the installation of panic hardware for means of egress doors equipped with locking devices?

A: Yes. Section 1008.1.9 requires panic hardware to be provided on egress doors that serve high occupant load densities in order to ensure prompt opening of the door in an emergency. The panic hardware provisions of Section

1008.1.9 apply to required egress doors that have all of the following characteristics:

1. Are equipped with latching or locking devices;
2. Serve an occupant load of 50 or more; and
3. Serve an assembly or educational occupancy.

Q: Is a door that is required to be equipped with panic hardware permitted to be equipped with a deadbolt lock that operates independently of the panic hardware?

A: No. Section 1008.1.8 requires all egress doors to be readily openable from the side from which egress is made without the use of a key, special knowledge or special effort. If the door is provided with a locking device that is not capable of being unlocked by panic hardware, that door would clearly be unavailable for egress when the locking device is engaged. Due to the uncertainty that the locking device will be unlocked during all times of occupancy, panic hardware must release all latching and locking devices on the required egress door.

[B] SECTION 1009 STAIRWAYS

[B] 1009.2 Headroom. Stairways shall have a minimum headroom clearance of 80 inches (2032 mm) measured vertically from a line connecting the edge of the nosings. Such headroom shall be continuous above the stairway to the point where the line intersects the landing below, one tread depth beyond the bottom riser. The minimum clearance shall be maintained the full width of the stairway and landing.

Exception: Spiral stairways complying with Section 1009.8 are permitted a 78-inch (1981 mm) headroom clearance.

Q: How is the minimum required headroom intended to be measured within all parts of a stairway in accordance with Section 1009.2?

A: The intent of the *International Building Code* regarding stairway headroom can best be described by considering a sloped reference line connecting the tread nosings. The minimum required headroom is represented by a sloped line parallel to and located 6 feet 8 inches vertically above the reference line for conventional stairways and 6 feet 6 inches vertically for spiral stairways.

[B] 1009.4 Stairway landings. There shall be a floor or landing at the top and bottom of each stairway. The width of landings shall not be less than the width of stairways they serve. Every landing shall have a minimum dimension measured in the direction of travel equal to the width of the stairway. Such dimension need not exceed 48 inches (1219 mm) where the stairway has a straight run.

Exceptions:

1. Aisle stairs complying with Section 1025.
2. Doors opening onto a landing shall not reduce the landing to less than one-half the required width. When fully open, the door shall not project more than 7 inches (178 mm) into a landing.

Q: Does the *International Building Code* permit a stairway landing to be shaped in a half circle rather than the typical rectangular-shaped area?

A: Yes. Section 1009.4 requires that the least dimension of a landing shall not be less than the required width of a stairway. If the landing provides an area described by an arc with a radius equal to the required stairway width, the minimum required space necessary for means of egress will be available subject to the restrictions of Section 1009.4.

[B] SECTION 1011 EXIT SIGNS

[B] 1011.2 Illumination. Exit signs shall be internally or externally illuminated.

Exception: Tactile signs required by Section 1011.3 need not be provided with illumination.

Q: Is it the intent of Section 1011.2 that all exit signs, including supplemental directional signs, be illuminated?

A: Yes. Section 1011.2 clearly intends that all exit signs be illuminated. The *International Building Code* requires that "all required means of egress" be indicated with approved illuminated exit signs. This encompasses both the exit access and the exit. The purpose of the supplemental directional exit sign is to direct building occupants to the exit if it is not readily visible from the exit access. Therefore, it is just as important for directional signs to be illuminated as it is for the exit sign itself to be illuminated.

[B] SECTION 1012 HANDRAILS

[B] 1012.4 Continuity. Handrail-gripping surfaces shall be continuous, without interruption by newel posts or other obstructions.

Exceptions:

1. Handrails within dwelling units are permitted to be interrupted by a newel post at a stair landing.
2. Within a dwelling unit, the use of a volute, turnout or starting easing is allowed on the lowest tread.
3. Handrail brackets or balusters attached to the bottom surface of the handrail that do not project horizontally beyond the sides of the handrail within 1.5 inches (38 mm) of the bottom of the handrail shall not be considered to be obstructions and provided further that for each 0.5 inch (13 mm) of additional handrail perimeter dimension above 4 inches (102 mm), the vertical clearance dimension of 1.5 inches (38 mm) shall be permitted to be reduced by 0.125 inch (3 mm).

Q: Are stairway handrails required to be constructed without obstructions, such as intermediate supports and posts?

A: Yes. Section 1012.4 requires continuous handrails to facilitate a reasonable degree of safety and assistance whenever ascending or descending the stairway. Obstructions that would significantly impair the ready use of the handrail gripping surfaces or that by configuration may cause injury through sharp or abrasive elements to individuals using the handrails are specifically prohibited.

Section 1002 defines "Stairway" as including the "necessary landings and platforms connecting them." Section 1012.4 states that stairways shall have continuous handrails on both sides.

Q: Are handrails required to extend around landings?

A: No. The code does not intend for handrails to be continuous around landings. Handrails are required to provide a horizontal or sloping rail for the occupant to grasp while negotiating the treads and risers. Section 1012.5 provides specific direction for handrail detailing at each landing.

[B] SECTION 1013 GUARDS

[B] 1013.1 Where required. Guards shall be located along open-sided walking surfaces, mezzanines, industrial equipment platforms, stairways, ramps and landings that are located more than 30 inches (762 mm) above the floor or grade below. Guards shall be adequate in strength and attachment in accordance with Section 1607.7. Where glass is used to provide a guard or as a portion of the guard system, the guard shall also comply with Section 2407. Guards shall also be located along glazed sides of stairways, ramps and landings that are located more than 30 inches (762 mm) above the floor or grade below where the glazing provided does not meet the strength and attachment requirements in Section 1607.7.

Exception: Guards are not required for the following locations:

1. On the loading side of loading docks or piers.
2. On the audience side of stages and raised platforms, including steps leading up to the stage and raised platforms.
3. On raised stage and platform floor areas, such as runways, ramps and side stages used for entertainment or presentations.
4. At vertical openings in the performance area of stages and platforms.
5. At elevated walking surfaces appurtenant to stages and platforms for access to and utilization of special lighting or equipment.
6. Along vehicle service pits not accessible to the public.
7. In assembly seating where guards in accordance with Section 1025.14 are permitted and provided.

Q: Is it the intent of Section 1013.1 to regulate an open-sided walking surface that is not a required element of a means of egress?

A: No. Section 1013.1 only applies to areas that are a required element of a means of egress. Most open-sided walking surfaces, mezzanines and landings are likely to be an element of the exit access, the exit or the exit discharge. If, however, there is an open-sided walking surface that is not an element of a means of egress, it can be evaluated by the code official as a potentially unsafe condition in accordance with Section 115.

[B] SECTION 1015
EXIT AND EXIT ACCESS DOORWAYS

[B] 1015.1 Exit or exit access doorways required. Two exits or exit access doorways from any space shall be provided where one of the following conditions exists:

1. The occupant load of the space exceeds the values in Table 1015.1.
2. The common path of egress travel exceeds the limitations of Section 1014.3.
3. Where required by Sections 1015.3, 1015.4 and 1015.5.

Exception: Group I-2 occupancies shall comply with Section 1014.2.2.

[B] TABLE 1015.1
SPACES WITH ONE MEANS OF EGRESS

OCCUPANCY	MAXIMUM OCCUPANT LOAD
A, B, E[a], F, M, U	49
H-1, H-2, H-3	3
H-4, H-5, I-1, I-3, I-4, R	10
S	29

a. Daycare maximum occupant load is 10.

Q: Do the maximum limitations in Table 1015.1 for buildings with one exit apply only to the portion of the building served by the single exit? For example, a building has two or more areas; each of which individually meets the limitations set forth in Table 1015.1 and has access to one exit. These areas are not separated from each other by fire walls.

A: No. The intent of Table 1015.1 is to limit the overall building size and occupant load in which occupants have access to only one exit. The building described exceeds, as a whole, the limitations of Table 1015.1 and provides all occupants with access to only one exit. In order for this arrangement to comply with Section 1015.1, each area must be separated from all others by a fire wall so that each area can be considered a separate building.

[B] 1015.2.1 Two exits or exit access doorways. Where two exits or exit access doorways are required from any portion of the exit access, the exit doors or exit access doorways shall be placed a distance apart equal to not less than one-half of the length of the maximum overall diagonal dimension of the building or area to be served measured in a straight line between exit doors or exit access doorways. Interlocking or scissor stairs shall be counted as one exit stairway.

Exceptions:

1. Where exit enclosures are provided as a portion of the required exit and are interconnected by a 1-hour fire-resistance-rated corridor conforming to the requirements of Section 1017, the required exit separation shall be measured along the shortest direct line of travel within the corridor.
2. Where a building is equipped throughout with an automatic sprinkler system in accordance with Section 903.3.1.1 or 903.3.1.2, the separation distance of the exit doors or exit access doorways shall not be less than one-third of the length of the maximum overall diagonal dimension of the area served.

Q: Does the *International Building Code* permit the use of scissor stairs as exits?

A: Yes. The use of scissor type stairs is not prohibited in Chapter 10. However, scissor stairs shall be counted as one exit stairway.

[B] SECTION 1016
EXIT ACCESS TRAVEL DISTANCE

[B] 1016.1 Travel distance limitations. Exits shall be so located on each story such that the maximum length of exit access travel, measured from the most remote point within a story to the entrance to an exit along the natural and unobstructed path of egress travel, shall not exceed the distances given in Table 1016.1.

Where the path of exit access includes unenclosed stairways or ramps within the exit access or includes unenclosed exit ramps or stairways as permitted in Section 1020.1, the distance of travel on such means of egress components shall also be included in the travel distance measurement. The measurement along stairways shall be made on a plane parallel and tangent to the stair tread nosings in the center of the stairway.

Exceptions:

1. Travel distance in open parking garages is permitted to be measured to the closest riser of open stairs.
2. In outdoor facilities with open exit access components and open exterior stairs or ramps, travel distance is permitted to be measured to the closest riser of a stair or the closest slope of the ramp.
3. Where an exit stair is permitted to be unenclosed in accordance with Exception 8 or 9 of Section 1020.1, the travel distance shall be measured from the most remote point within a building to an exit discharge.

[B] TABLE 1016.1
EXIT ACCESS TRAVEL DISTANCE[a]

OCCUPANCY	WITHOUT SPRINKLER SYSTEM (feet)	WITH SPRINKLER SYSTEM (feet)
A, E, F-1, I-1, M, R, S-1	200	250[b]
B	200	300[c]
F-2, S-2, U	300	400[b]
H-1	Not Permitted	75[c]
H-2	Not Permitted	100[c]
H-3	Not Permitted	150[c]
H-4	Not Permitted	175[c]
H-5	Not Permitted	200[c]
I-2, I-3, I-4	150	200[c]

For SI: 1 foot = 304.8 mm.

a. See the following sections for modifications to exit access travel distance requirements:
Section 402: For the distance limitation in malls.
Section 404: For the distance limitation through an atrium space.
Section 1015.2: For increased limitation in Groups F-1 and S-1.
Section 1024.7: For increased limitation in assembly seating.
Section 1024.7: For increased limitation for assembly open-air seating.
Section 1018.2: For buildings with one exit.
Chapter 31: For the limitation in temporary structures.

b. Buildings equipped throughout with an automatic sprinkler system in accordance with Section 903.3.1.1 or 903.3.1.2. See Section 903 for occupancies where sprinkler systems according to Section 903.3.1.2 are permitted.

c. Buildings equipped throughout with an automatic sprinkler system in accordance with Section 903.3.1.1.

Q: An enclosed mezzanine (in compliance with Section 505.4 Exception 2 of the *International Building Code*) is provided with one exit stairway that discharges into a sprinklered lobby in accordance with Section 1024.1 and one unenclosed stairway. Are the maximum travel distance limitations of Section 1016 applicable to the means of egress from the enclosed mezzanine?

A: Yes. The maximum exit access travel distance identified in Table 1016.1 is measured from each and every occupiable point within the building (including the mezzanine level) to an exit. Although the occupants may be required by Section 1019.1 to have access to a second, third or fourth exit, only the distance to the nearest exit is regulated by the code. Therefore, the distance an occupant must travel to reach a second, third or fourth required exit is not limited.

The occupants of the mezzanine described above are provided with two means of egress. The first exit access path begins at the most remote point of the mezzanine, proceeds across the mezzanine to the unenclosed stairway, down the stairway and terminates at the exit serving the main floor. The second exit access path also begins at the most remote point of the mezzanine, proceeds across the mezzanine and terminates at the entrance of the enclosed exit stairway leading from the mezzanine level to the exit lobby. Traveling along one of these exit access paths, occupants of the mezzanine must reach the safety of an exit within the maximum travel distance identified in Section 1016. The distance traveled within the safety of exit elements, such as the enclosed exit stairway and the exit lobby, is not limited.

[B] SECTION 1017
CORRIDORS

[B] 1017.1 Construction. Corridors shall be fire-resistance rated in accordance with Table 1017.1. The corridor walls required to be fire-resistance rated shall comply with Section 708 for fire partitions.

Exceptions:

1. A fire-resistance rating is not required for corridors in an occupancy in Group E where each room that is used for instruction has at least one door directly to the exterior and rooms for assembly purposes have at least one-half of the required means of egress doors opening directly to the exterior. Exterior doors specified in this exception are required to be at ground level.
2. A fire-resistance rating is not required for corridors contained within a dwelling or sleeping unit in an occupancy in Group R.
3. A fire-resistance rating is not required for corridors in open parking garages.
4. A fire-resistance rating is not required for corridors in an occupancy in Group B which is a space requiring only a single means of egress complying with Section 1015.1.

Q: Does Section 1017 require exit access corridors to be provided in buildings?

A: No. Section 1017 only provides requirements for exit access corridors whenever they are provided in a building. The *International Building Code* requires access to exits for every building, which can be accomplished by open spaces, aisles or corridors. A corridor is created when the path of travel to an exit is confined, and communication with and observation of adjacent floor areas are restricted.

Q: A floor plan, of which a portion is an open plan without corridors and an adjacent portion that is compartmented, contains a corridor that leads to the open plan area. Assuming the corridor enclosure is required to be fire-resistance rated, is it the intent of the code to require that an opening protective be installed at the point where the corridor enters the open area?

A: No. The relatively open areas adjacent to the corridor can be considered corridor extensions with rated walls and opening protectives enclosing both the narrow and wide portions of the corridor in a continuous fashion.

[B] 1017.2 Corridor width. The minimum corridor width shall be as determined in Section 1005.1, but not less than 44 inches (1118 mm).

Exceptions:

1. Twenty-four inches (610 mm)—For access to and utilization of electrical, mechanical or plumbing systems or equipment.
2. Thirty-six inches (914 mm)—With a required occupant capacity of less than 50.
3. Thirty-six inches (914 mm)—Within a dwelling unit.
4. Seventy-two inches (1829 mm)—In Group E with a corridor having a required capacity of 100 or more.
5. Seventy-two inches (1829 mm)—In corridors serving surgical Group I, health care centers for ambulatory patients receiving outpatient medical care, which causes the patient to be not capable of self-preservation.
6. Ninety-six inches (2438 mm)—In Group I-2 in areas where required for bed movement.

Q: Section 1017.2 indicates that the required width of corridors can be reduced to 36 inches where serving an occupant load of less than 50. In layouts where dead-end or cross aisles and corridors serve limited portions of a floor area, can such passageways, aisle accessways, aisles and corridors be analyzed individually in segments so that only the occupant load that would actually traverse the exit access element to reach a required exit be used to determine the required width?

A: Yes. Certain passageway, aisle and corridor arrangements can be readily evaluated to determine the actual occupant load served by such element. In these cases, it can often be recognized that a given segment does not actually serve the entire occupant load of the room. In such cases, only the portion of the occupant load that would traverse part of a passageway, aisle or corridor to reach an exit need be used to determine the required width.

Q: Does the 72-inch minimum required corridor width apply to every corridor in a Group E building having an occupant load greater than 100?

A: No. Item 4 of Section 1017.2 is intended to apply only to corridors, aisle accessways, aisles or passageways serving 100 or more occupants. Corridors serving 51 to 99 occupants are required to be at least 44 inches wide, whereas corridors serving 50 or fewer occupants are required to be at least 36 inches wide.

[B] 1017.3 Dead ends. Where more than one exit or exit access doorway is required, the exit access shall be arranged such that there are no dead ends in corridors more than 20 feet (6096 mm) in length.

Exceptions:

1. In occupancies in Group I-3 of Occupancy Condition 2, 3 or 4 (see Section 308.4), the dead end in a corridor shall not exceed 50 feet (15 240 mm).
2. In occupancies in Groups B and F where the building is equipped throughout with an automatic sprinkler system in accordance with Section 903.3.1.1, the length of dead-end corridors shall not exceed 50 feet (15 240 mm).
3. A dead-end corridor shall not be limited in length where the length of the dead-end corridor is less than 2.5 times the least width of the dead-end corridor.

Q: A door is installed in an exit access corridor serving two exits, and the door only swings in one direction. Does the installation of this door constitute a dead-end corridor for the occupants traveling against the door swing?

A: No. A dead-end corridor is established when occupants are required to retrace their steps to gain access to an exit. A doorway across a corridor does not block the path of travel in the corridor so that occupants have to retrace their steps to reach an exit. However, Section 1008.1.2 establishes when a door must swing in the direction of travel. If the doorway in question serves either an occupant load of 50 or more in each direction or a Group H occupancy, the door must be capable of swinging in both directions or two doors, each swinging in opposite directions, must be provided.

Q: When more than one exit is required by the code, Section 1019.1 requires occupants to have direct access to the exits in separate directions. Section 1017.3 appears to conflict with Section 1019.1 because a dead-end corridor does not provide access to the exits in separate directions. Is it the intent of these sections to permit a dead-end corridor to exist in a building required to have more than one exit?

A: Yes. The requirements of Section 1019.1 are intended to provide exits that are remote as practicable from each other in accordance with Section 1015.2.1. Dead-end exit access travel within the limits of Section 1017.3 is specifically intended to be permitted and represents the limit at which the exits would still be deemed to be as remote as practicable.

[B] 1017.4 Air movement in corridors. Corridors shall not serve as supply, return, exhaust, relief or ventilation air ducts.

Exceptions:

1. Use of a corridor as a source of makeup air for exhaust systems in rooms that open directly onto such corridors, including toilet rooms, bathrooms, dressing rooms, smoking lounges and janitor closets, shall be permitted, provided that each such corridor is directly supplied with outdoor air at a rate greater than the rate of makeup air taken from the corridor.
2. Where located within a dwelling unit, the use of corridors for conveying return air shall not be prohibited.
3. Where located within tenant spaces of 1,000 square feet (93 m^2) or less in area, utilization of corridors for conveying return air is permitted.

Q: Can a single HVAC duct system (supply, return, exhaust, relief, or ventilation) serve an exit access corridor and spaces open and/or adjacent to the exit access corridor where the air supplied into and returned out of the open and/or adjacent spaces is through the duct system and not through the exit access corridor?

A: Yes. Air supplied from a single HVAC duct system is permitted to serve an exit access corridor and spaces open and/or adjacent to the exit access corridor. Air supplied by the duct system into the spaces open and/or adjacent to the exit access corridor, either by the duct system or by other mechanical or nonmechanical means, is not permitted to be transferred into or through the exit access corridor as supply air, return air, make up air, ventilation or exhaust.

[B] SECTION 1019 NUMBER OF EXIT AND CONTINUITY

[B] 1019.1 Minimum number of exits. All rooms and spaces within each story shall be provided with and have access to the minimum number of approved independent exits required by Table 1019.1 based on the occupant load of the story, except as modified in Section 1015.1 or 1019.2. For the purposes of this chapter, occupied roofs shall be provided with exits as required for stories. The required number of exits from any story, basement or individual space shall be maintained until arrival at grade or the public way.

Q: In a building required to have a minimum of two exits, can a stairway serve as a "pass-through" space for a corridor continuing beyond the stairway and leading to a second exit?

A: No. Section 1019.1 requires the exits to be independent. A means of egress arrangement requiring building occupants to enter or use one exit as the only means to gain access to a second exit does not provide independent exits.

Q: In a building in which two or more exits are required by Section 1019.1, can a room or tenant space provided with only one means of egress doorway, in accordance with Section 1015.1, discharge directly into one of the required exits?

A: No. Section 1019.1 states that every floor area shall be provided with the minimum number of approved independent exits. In the condition described above, the alternative exit is not independent of the exit into which the room or tenant space discharges. Access to the alternative exit depends on the primary exit remaining available for use during an emergency. If the primary exit is rendered unusable, the secondary exit will also be unavailable, effectively eliminating the independence of the alternative exit. The *International Building Code* would allow an egress door from a room or space to discharge directly into an exit, provided there is a second egress door from the room or space offering independent access to the other required exit(s).

This situation should not be confused with the issue presented by a lobby used as an exit element in accordance with Section 1024.1. In this case, the lobby serves as an exit element for the occupants located only on the upper floors. For occupants located on the lobby floor, however, the lobby serves as an exit access passageway and can be used to provide independent access to the required exits (exterior doors).

Q: Are the minimum number of exits regulated by Section 1019.1 required to be available to all floor area occupants?

A: Yes. The basic concepts regarding means of egress require that approved independent exits must be provided from every floor area of a building. The minimum number of exits serving every floor area is determined in accordance with Section 1019.1 based on the occupant load of the floor area in question. These exits can be any of the types permitted by the code. An unenclosed stairway, other than as allowed by the exceptions to Section 1020.1 and other than those serving mezzanines and penthouses, can only provide convenient or supplemental access to an adjacent floor, and must comply with the provisions of either Section 404 or 707.2. An egress arrangement would not be permitted wherein an occupant must travel to an adjacent story through such a stairway to gain access to a required exit.

[B] 1019.2 Buildings with one exit. Only one exit shall be required in buildings as described below:

1. Buildings described in Table 1019.2, provided that the building has not more than one level below the first story above grade plane.
2. Buildings of Group R-3 occupancy.
3. Single-level buildings with the occupied space at the level of exit discharge provided that the story or space complies with Section 1015.1 as a space with one means of egress.

[B]TABLE 1019.2
BUILDINGS WITH ONE EXIT

OCCUPANCY	MAXIMUM HEIGHT OF BUILDING ABOVE GRADE PLANE	MAXIMUM OCCUPANTS (OR DWELLING UNITS) PER FLOOR AND TRAVEL DISTANCE
A, B[d], E[e], F, M, U	1 Story	49 occupants and 75 feet travel distance
H-2, H-3	1 Story	3 occupants and 25 feet travel distance
H-4, H-5, I, R	1 Story	10 occupants and 75 feet travel distance
S[a]	1 Story	29 occupants and 100 feet travel distance
B[b], F, M, S[a]	2 Stories	30 occupants and 75 feet travel distance
R-2	2 Stories[c]	4 dwelling units and 50 feet travel distance

For SI: 1 foot = 304.8 mm.

a. For the required number of exits for open parking structures, see Section 1019.1.1.

b. For the required number of exits for air traffic control towers, see Section 412.1.

c. Buildings classified as Group R-2 equipped throughout with an automatic sprinkler system in accordance with Section 903.3.1.1 or 903.3.1.2 and provided with emergency escape and rescue openings in accordance with Section 1026 shall have a maximum height of three stories above grade plane.

d. Buildings equipped throughout with an automatic sprinkler system in accordance with Section 903.3.1.1 with an occupancy in Group B shall have a maximum travel distance of 100 feet.

e. Day care maximum occupant load is 10.

Q: Does a basement story that is not a story above grade but contains an exterior "areaway" stairway leading from that story constitute the level of exit discharge?

A: No. Below-grade or partially below-grade stories are not at the level of exit discharge, just as a second story above grade is not at the level of exit discharge. In this instance, the exit discharge is located at the top of the exterior "areaway" stairway (at the grade elevation). Thus, a level below the level of exit discharge is required to have a minimum of two approved exits unless the building qualifies for a single exit.

[B] SECTION 1020
VERTICAL EXIT ENCLOSURES

[B] 1020.1 Enclosures required. Interior exit stairways and interior exit ramps shall be enclosed with fire barriers constructed in accordance with Section 706 of the *International Building Code* or horizontal assemblies constructed in accordance with Section 711 of the *International Building Code*, or both. Exit enclosures shall have a fire-resistance rating of not less than 2 hours where connecting four stories or more and not less than 1 hour where connecting less than four stories. The number of stories connected by the exit enclosure shall include any basements but not any mezzanines. An exit enclosure shall not be used for any purpose other than means of egress.

Exceptions:

1. In all occupancies, other than Group H and I occupancies, a stairway is not required to be enclosed when the stairway serves an occupant load of less than 10 and the stairway complies with either Item 1.1 or 1.2. In all cases, the maximum number of connecting open stories shall not exceed two.
   1.1. The stairway is open to not more than one story above the story at the level of exit discharge; or
   1.2. The stairway is open to not more than one story below the story at the level of exit discharge.
2. Exits in buildings of Group A-5 where all portions of the means of egress are essentially open to the outside need not be enclosed.
3. Stairways serving and contained within a single residential dwelling unit or sleeping unit in Group R-1, R-2 or R-3 occupancies are not required to be enclosed.
4. Stairways that are not a required means of egress element are not required to be enclosed where such stairways comply with Section 707.2.
5. Stairways in open parking structures that serve only the parking structure are not required to be enclosed.
6. Stairways in Group I-3 occupancies, as provided for in Section 408.3.6, are not required to be enclosed.
7. Means of egress stairways as required by Section 410.5.3 are not required to be enclosed.
8. In other than Group H and I occupancies, a maximum of 50 percent of egress stairways serving one adjacent floor are not required to be enclosed, provided at least two means of egress are provided from both floors served by the unenclosed stairways. Any two such interconnected floors shall not be open to other floors. Unenclosed exit stairways shall be remotely located as required in Section 1015.2.
9. In other than Group H and I occupancies, interior egress stairways serving only the first and second stories of a building equipped throughout with an automatic sprinkler system in accordance with Section

903.3.1.1 are not required to be enclosed, provided at least two means of egress are provided from both floors served by the unenclosed stairways. Such interconnected stories shall not be open to other stories. Unenclosed exit stairways shall be remotely located as required in Section 1015.2.

Q: Given a building on grade, without a basement, see Attachment No. 1 Multi-Story Building. Is the second story of this building the first story above the lowest level of exit discharge?

A: No. When determining stories above the lowest level of exit discharge, a level, or floor level, is not a story. A "level" is the horizontal plane that is part of a story, not the entire story height. A "story" is the vertical space between the upper surface of one floor level and the upper surface of the floor level next above or below.

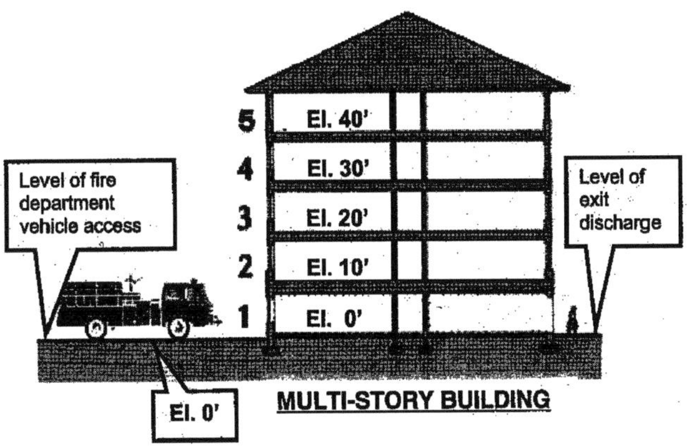

Attachment No. 1

The level of exit discharge as shown in attached drawing of a Multi-Story Building is at elevation 0.0 feet, the first floor level of the building is also at elevation 0.0 feet; therefore, the level of exit discharge and the first floor level of the building are at the same elevation. The first story of the building begins at elevation 0.0 feet (first floor) and extends to elevation 10.0 feet (second floor). The first story of the building is the first story above the level of exit discharge.

Q: Are scissor stairs (e.g., two flights of stairs) allowed in the same enclosure under Section 1020.1?

A: Yes. Section 1020.1 does not limit the number of stairways allowed within a single enclosure. However, if more than one stairway is located in a single enclosure without being completely separated by a fire separation assembly with a fire-resistance rating as required for exits, the stairways would only be considered one exit for the purpose of the required number of exits.

[B] 1020.1.1 Openings and penetrations. Exit enclosure opening protectives shall be in accordance with the requirements of Section 715 of the *International Building Code.*

Except as permitted in Section 402.4.6 of the *International Building Code*, openings in exit enclosures other than unprotected exterior openings shall be limited to those necessary for exit access to the enclosure from normally occupied spaces and for egress from the enclosure.

Where interior exit enclosures are extended to the exterior of a building by an exit passageway, the door assembly from the exit enclosure to the exit passageway shall be protected by a fire door assembly conforming to the requirements in Section 715.4 of the *International Building Code.* Fire door assemblies in exit enclosures shall comply with Section 715.4.4 of the *International Building Code.*

Elevators shall not open into an exit enclosure.

Q: Are all doors, except corridor doors, prohibited by Section 1020.1.1 from opening into an exit stairway?

A: No. The intent of Section 1020.1.1 is to permit only doors that are part of an element of a means of egress to open into stairway enclosures. Rooms such as janitor's closets, mechanical equipment rooms, closets and other normally unoccupied spaces are prohibited from having doors opening into the stairway enclosure. Thus, rooms that would usually be occupied can have doors opening directly into the stairway enclosure. For buildings in which two or more exits are required, the means of egress from such occupied rooms must provide access to the other required exits in a manner that does not require passing through the exit stairway in question.

Q: An automotive service shop has below-grade service areas where employees perform oil changes and other minor maintenance services. The below-grade areas are open to the grade-level service bays, thus providing access to the underside of the vehicles without requiring them to be lifted into the air. A stairway is provided to gain access to and egress from this below-grade service area. Is the stairway from this below-grade service level required to be enclosed in fire-resistance-rated construction in accordance with Section 1020.1?

A: No. Section 1020.1 requires exit stairways that connect two or more stories to be enclosed in fire-resistance-rated construction. The enclosure requirements would only apply to this below-grade service level if it is, in fact, a story. Due to the openness between the adjacent service levels, the below-grade area would be more appropriately regulated similar to a mezzanine rather than a story. A mezzanine is considered part of the same story that it serves, and egress stairways from the mezzanine level are not required to be enclosed in fire-resistance-rated construction in accordance with Exception 9 to Section 707.2.

Therefore, if the below-grade service level is in compliance with the applicable provisions of Section 505, the egress stairway serving the mezzanine level is not required to be enclosed in fire-resistance-rated construction.

[B] 1020.1.2 Penetrations. Penetrations into and openings through an exit enclosure are prohibited except for required exit doors, equipment and ductwork necessary for independent pressurization, sprinkler piping, standpipes, electrical raceway for fire department communication and electrical raceway serving the exit enclosure and terminating at a steel box not exceeding 16 square inches (0.010 m^2). Such penetrations shall be protected in accordance with Section 712. There shall be no penetrations or communication openings, whether protected or not, between adjacent exit enclosures.

Q: Is a hot water unit heater permitted to be installed within an exit enclosure?

A: Yes. Portions of a buildings service systems that serve only the exit enclosure are permitted to penetrate the exit enclosure that is being served. The hot water unit heater cannot obstruct the path of egress travel, or reduce the clear width of the stair or landing. The boiler that provides the hot water for the unit heater must be located outside the exit enclosure.

Q: Is it the intent of Section 1020.1.2 to prohibit hose connections on standpipes from being located in the stairway enclosure?

A: No. The code provision requiring an exit stairway enclosure to not be used for any purpose other than means of egress is intended to prohibit hazards to means of egress and also to restrict the possibility of locating a fuel load within the enclosure. Use of the enclosure as a "staging area" by a fire department does not represent "use" of a stairway enclosure intended to be restricted by this section. Furthermore, NFPA 14, the standard referenced by the code, specifically requires, with certain exceptions, that standpipe risers be located within fire-resistance-rated stairway enclosures.

[B] 1020.1.3 Ventilation. Equipment and ductwork for exit enclosure ventilation as permitted by Section 1020.1.2 shall comply with one of the following items:

1. Such equipment and ductwork shall be located exterior to the building and shall be directly connected to the exit enclosure by ductwork enclosed in construction as required for shafts.
2. Where such equipment and ductwork is located within the exit enclosure, the intake air shall be taken directly from the outdoors and the exhaust air shall be discharged directly to the outdoors, or such air shall be conveyed through ducts enclosed in construction as required for shafts.
3. Where located within the building, such equipment and ductwork shall be separated from the remainder of the building, including other mechanical equipment, with construction as required for shafts.

In each case, openings into the fire-resistance-rated construction shall be limited to those needed for maintenance and operation and shall be protected by opening protectives in accordance with Section 715 for shaft enclosures.

Exit enclosure ventilation systems shall be independent of other building ventilation systems.

Q: Are ducts permitted to penetrate the fire-resistance-rated enclosure walls of an exit stairway for purposes of heating, ventilating and cooling the stair enclosure?

A: Yes. Section 716.5.3 allows ducts to service the stairway enclosure when permitted by Section 1020.1.2. (Note: Section 1020.1.1 restricts penetrations to only those serving the exit itself.)

[B] SECTION 1023
EXTERIOR EXIT RAMPS AND STAIRWAYS

[B] 1023.6 Exterior ramps and stairway protection. Exterior exit ramps and stairways shall be separated from the interior of the building as required in Section 1020.1. Openings shall be limited to those necessary for egress from normally occupied spaces.

Exceptions:

1. Separation from the interior of the building is not required for occupancies, other than those in Group R-1 or R-2, in buildings that are no more than two stories above grade plane where the level of exit discharge is the first story above grade plane.
2. Separation from the interior of the building is not required where the exterior ramp or stairway is served by an exterior ramp and/or balcony that connects two remote exterior stairways or other approved exits with a perimeter that is not less than 50 percent open. To be considered open, the opening shall be a minimum of 50 percent of the height of the enclosing wall, with the top of the openings no less than 7 feet (2134 mm) above the top of the balcony.

3. Separation from the interior of the building is not required for an exterior ramp or stairway located in a building or structure that is permitted to have unenclosed interior stairways in accordance with Section 1020.1.
4. Separation from the interior of the building is not required for exterior ramps or stairways connected to open-ended corridors, provided that Items 4.1 through 4.4 are met:
    - 4.1. The building, including corridors and ramps and/or stairs, shall be equipped throughout with an automatic sprinkler system in accordance with Section 903.3.1.1 or 903.3.1.2.
    - 4.2. The open-ended corridors comply with Section 1017.
    - 4.3. The open-ended corridors are connected on each end to an exterior exit ramp or stairway complying with Section 1023.
    - 4.4. At any location in an open-ended corridor where a change of direction exceeding 45 degrees (0.79 rad) occurs, a clear opening of not less than 35 square feet (3.3 m^2) or an exterior ramp or stairway shall be provided. Where clear openings are provided, they shall be located so as to minimize the accumulation of smoke or toxic gases.

Q: Are exterior exit stairs required to be separated with fire-resistive-rated construction from open-ended corridors?

A: No. Exception 4 of Section 1023.6 defines specific provisions that permit the use of exterior stairs connected to open-ended corridors. An exterior stairway need not be separated from the interior of the building or from the open-ended corridor where both comply with the provisions in Items 4.1 through 4.4.

[B] SECTION 1024
EXIT DISCHARGE

[B] 1024.1 General. Exits shall discharge directly to the exterior of the building. The exit discharge shall be at grade or shall provide direct access to grade. The exit discharge shall not reenter a building.

Exceptions:

1. A maximum of 50 percent of the number and capacity of the exit enclosures is permitted to egress through areas on the level of discharge provided all of the following are met:
    - 1.1. Such exit enclosures egress to a free and unobstructed way to the exterior of the building, which way is readily visible and identifiable from the point of termination of the exit enclosure.
    - 1.2. The entire area of the level of discharge is separated from areas below by construction conforming to the fire-resistance rating for the exit enclosure.
    - 1.3. The egress path from the exit enclosure on the level of discharge is protected throughout by an approved automatic sprinkler system. All portions of the level of discharge with access to the egress path shall either be protected throughout with an automatic sprinkler system installed in accordance with Section 903.3.1.1 or 903.3.1.2, or separated from the egress path in accordance with the requirements for the enclosure of exits.
2. A maximum of 50 percent of the number and capacity of the exit enclosures is permitted to egress through a vestibule provided all of the following are met:
    - 2.1. The entire area of the vestibule is separated from areas below by construction conforming to the fire-resistance rating for the exit enclosure.
    - 2.2. The depth from the exterior of the building is not greater than 10 feet (3048 mm) and the length is not greater than 30 feet (9144 mm).
    - 2.3. The area is separated from the remainder of the level of exit discharge by construction providing protection at least the equivalent of approved wired glass in steel frames.
    - 2.4. The area is used only for means of egress and exits directly to the outside.
3. Stairways in open parking garages complying with Section 1020.1, Exception 5, are permitted to egress through the open parking garage at the level of exit discharge.

Q: Is the "required discharge capacity" of an exit discharge, as stated in Section 1024.2, to be determined using Section 1004.1?

A: Yes. The exit discharge is one component of a means of egress and must be sized to carry adequately the occupant load served. Table 1004.1.1 would be applied to determine that the width required to provide adequate capacity is achieved. Corridor and stairway width in accordance with Section 1005.1 would also be used to determine that the minimum width of the exit discharge is achieved.

Q: An enclosed mezzanine with more than ten occupants is provided with one enclosed exit stairway and one unenclosed stairway in accordance with Exception 2 of Section 505.4. Does Section 1024.1 prohibit both means of egress stairways from discharging into the same space at the floor below where the space qualifies as a lobby in accordance with Section 1024.1, Exception 1?

A: No. The provisions of Section 1024.1 are applicable to multiple exit elements that discharge into the same space within the building. An unenclosed stairway from the mezzanine is not considered an exit, but rather an element of exit access. The unenclosed stairway is not regulated by Section 1024.1. Therefore, both means of egress stairways that serve the mezzanine are permitted to discharge into the common lobby.

Q: Does Section 1024.1 require that a public way meet the means of egress requirements in the code relative to slip resistance, levelness, curbs, steps, illumination, etc.?

A: No. The means of egress exit elements of a building or structure must discharge directly to a public way or to a space serving as the exit discharge which ultimately leads to a public way. While the code provides specific requirements for the exit discharge leading to the public way, the code does not identify specific requirements for a public way. The definition of "Public way" establishes that it is limited to public use and has a clear width and a clear height of at least 10 feet. All other attributes of a public way are outside the scope of the code.

The intent of the means of egress requirements leading to a public way is to provide the occupants with a safe area where they are free from the immediate effects of a fire or other emergency situation within the building or structure. The code does not contain requirements for slip resistance, levelness, curbs, steps, illumination, etc., for public streets, public alleys or open parcels of land legally limited to public use.

[B] SECTION 1026 EMERGENCY ESCAPE AND RESCUE

[B] 1026.1 General. In addition to the means of egress required by this chapter, provisions shall be made for emergency escape and rescue in Group R and I-1 occupancies. Basements and sleeping rooms below the fourth story above grade plane shall have at least one exterior emergency escape and rescue opening in accordance with this section. Where basements contain one or more sleeping rooms, emergency egress and rescue openings shall be required in each sleeping room, but shall not be required in adjoining areas of the basement. Such openings shall open directly into a public way or to a yard or court that opens to a public way.

Exceptions:

1. In other than Group R-3 occupancies, buildings equipped throughout with an approved automatic sprinkler system in accordance with Section 903.3.1.1 or 903.3.1.2.
2. In other than Group R-3 occupancies, sleeping rooms provided with a door to a fire-resistance-rated corridor having access to two remote exits in opposite directions.
3. The emergency escape and rescue opening is permitted to open onto a balcony within an atrium in accordance with the requirements of Section 404 of the *International Building Code*, provided the balcony provides access to an exit and the dwelling unit or sleeping unit has a means of egress that is not open to the atrium.
4. Basements with a ceiling height of less than 80 inches (2032 mm) shall not be required to have emergency escape and rescue windows.
5. High-rise buildings in accordance with Section 403 of the *International Building Code*.
6. Emergency escape and rescue openings are not required from basements or sleeping rooms that have an exit door or exit access door that opens directly into a public way or to a yard, court or exterior exit balcony that opens to a public way.
7. Basements without habitable spaces and having no more than 200 square feet (18.6 m^2) in floor area shall not be required to have emergency escape windows.

Q: A multistory building with an occupancy in Group R or I-1 has an exit access corridor and exit stairways separated from an atrium. If the sleeping rooms below the fourth story have windows of the minimum required dimensions and area opening onto the atrium, which then provides direct access to the outside, is the intent of Exception 3 to Section 1026.1 satisfied?

A: Yes. The purpose of the emergency escape window is to provide an alternative means of escape or rescue in the event that the required means of egress is not available. If the required means of egress from the sleeping room does not directly communicate with the atrium space, the likelihood of both routes blocked by the same incident is remote. On this basis, the intent of Exception 3 to Section 1026.1 has been satisfied.

Q: In a building with an occupancy in Group R or I-1, which is required to have a minimum of two exits per floor, would windows complying with the requirements of Section 1026.2 be considered acceptable as one of the required exits?

A: No. An emergency escape window is required in addition to the minimum number of required exits. The window is intended to serve the special function of emergency escape or rescue for sleeping rooms only and is not intended to serve as a substitution for the required exit facilities for a floor area.

[B] 1026.2 Minimum size. Emergency escape and rescue openings shall have a minimum net clear opening of 5.7 square feet (0.53 m^2).

> **Exception:** The minimum net clear opening for emergency escape and rescue grade-floor openings shall be 5 square feet (0.46 m^2).

Q: What constitutes "grade-floor" openings as specified in the exception to Section 1026.2?

A: The intent of the exception is to permit a reduction in the minimum net clear opening of 5.7 square feet to 5 square feet for those windows that can be accessed from the adjoining grade without the use of a ladder.

CHAPTER 23

HIGH-PILED COMBUSTIBLE STORAGE

SECTION 2302 DEFINITIONS

2302.1 Definitions. The following words and terms shall, for the purposes of this chapter and as used elsewhere in this code, have the meanings shown herein.

REFERENCED SECTION:

SECTION 2306 GENERAL LIFE PROTECTION AND LIFE SAFETY FEATURES

2306.9 Aisles. Aisles providing access to exits and fire department access doors shall be provided in high-piled storage areas exceeding 500 square feet (46 m^2), in accordance with Sections 2306.9.1 through 2306.9.3. Aisles separating storage piles or racks shall comply with NFPA 13. Aisles shall also comply with Chapter 10.

> **Exception:** Where aisles are precluded by rack storage systems, alternate methods of access and protection are allowed when approved.

HIGH-PILED STORAGE AREA. An area within a building which is designated, intended, proposed or actually used for high-piled combustible storage.

Q: Within a building, when calculating the actual floor area for that portion of the building dedicated for use as "High-piled storage area", as defined in Section 2302.1 of the *International Fire Code*, are aisles that are required for access to exits and fire department access doors considered part of the high-piled storage area?

A: No. Where aisles are provided in addition to the aisles required by NFPA 13 they need not be included in the high-piled storage area. Section 2306.9 requires aisles for both access to exits and fire department access doors in addition to aisles separating storage piles or racks. Only those aisles separating storage piles and racks that are required by NFPA 13 need to be included in the high-piled storage area.

2306.6.1.1 Number of doors required. A minimum of one access door shall be provided in each 100 lineal feet (30 480 mm), or fraction thereof, of the exterior walls which face required fire apparatus access roads.

Q: Where fire department access doors are required by Table 2306.2, Section 2306.6.1.1 requires the doors to be provided in each 100 lineal feet of exterior wall, or fraction thereof. Are the access doors required to be located such that the maximum distance between each door does not exceed 100 lineal feet?

A: Yes. The required openings must be distributed such that the lineal distance between adjacent openings does not exceed 100 feet.

CHAPTER 27

HAZARDOUS MATERIALS—GENERAL PROVISIONS

SECTION 2703
GENERAL REQUIREMENTS

2703.1.1 Maximum allowable quantity per control area. The maximum allowable quantity per control area shall be as specified in Tables 2703.1.1(1) through 2703.1.1(4).

For retail and wholesale storage and display in Group M occupancies and Group S storage, see Section 2703.11.

Q: Do the quantities of liquefied petroleum gas provided in Chapter 38 take precedence over the maximum allowable quantities per control area in Table 2703.1.1(1) for LP-gas with respect to determining occupancy classification?

A: No. Occupancy classifications are determined by the provisions in the *International Building Code*, which provides no exceptions to the applicability of [F]Table 307.7(1) Maximum Allowable Quantities per Control Area in determining occupancy classifications.

INTERNATIONAL CODE INTERPRETATIONS

MECHANICAL

2006

ICC

INTERNATIONAL CODE COUNCIL®

CHAPTER 3

GENERAL REGULATIONS

SECTION 304
INSTALLATION

304.3 Elevation of ignition source. Equipment and appliances having an ignition source and located in hazardous locations and public garages, private garages, repair garages, automotive motor-fuel-dispensing facilities and parking garages shall be elevated such that the source of ignition is not less than 18 inches (457 mm) above the floor surface on which the equipment or appliance rests. Such equipment and appliances shall not be installed in Group H occupancies or control areas where open use, handling or dispensing of combustible, flammable or explosive materials occurs. For the purpose of this section, rooms or spaces that are not part of the living space of a dwelling unit and that communicate directly with a private garage through openings shall be considered to be part of the private garage.

Q: A private garage attached to a dwelling unit contains a 4-inch-high platform installed over a small portion of the garage floor. The required water heater for the residence will be located upon the platform. Does Section 305.3 of the *International Fuel Gas Code* and Section 304.3 of the *International Mechanical Code* require the 18-inch installation height for the ignition source to be measured from the top of the platform?

A: No. A private garage not only contains vehicles with fuel tanks but also is commonly used as a storage area for containers of gasoline and other hazardous household items such as paint, thinners and solvents. These flammable and combustible liquids produce vapors that are more dense than air. As these vapors collect on the garage floor, they represent a serous fire hazard due to the ignition potential of fuel-burning water heaters, mechanical equipment and appliances.

To reduce substantially the risk of explosion and fire, Section 305.3 of the *International Fuel Gas Code* and Section 304.3 of the *International Mechanical Code* require that fuel-burning water heaters, mechanical equipment and appliances be elevated such that their ignition source is not less than 18 inches above the garage floor. In this case, the 18-inch installation height can be measured from the main garage floor where the flammable vapors will accumulate. Note this assumes the raised platform is limited in size such that it will not be used for the storage of flammable and combustible liquids.

304.7 Construction and protection. Boiler rooms and furnace rooms shall be protected as required by the *International Building Code*.

Q: Is a room enclosing a fuel-fired water heater in excess of 400,000 Btu per hour input considered a furnace room and regulated as a specific occupancy area in accordance with Table 508.2 of the *International Building Code?*

A: Yes. The definition of "Furnace room" in Section 202 of the *International Mechanical Code* includes rooms that enclose fuel-fired furnaces and similar fuel-fired equipment. Section 304.7 of the *International Mechanical Code* requires furnace rooms to be protected as required by the *International Building Code*. Such rooms are often utilized for unapproved storage, thus creating a fire hazard where combustible materials are stored in close proximity to the appliance. Although a water heater may or may not be utilized for space-heating purposes, it poses the same relative hazard as any other piece of fuel-fired heating equipment confined in a room. Thus, a room enclosing a fuel-fired water heater is regulated as a furnace room and subject to the specific occupancy requirements of Table 508.2 of the *International Building Code.*

CHAPTER 4

VENTILATION

SECTION 401 GENERAL

401.3 When required. Ventilation shall be provided during the periods that the room or space is occupied.

Q: Section 401.3 requires that ventilation be provided during periods when a room or space is occupied. Certain rooms requiring ventilation, such as public and private bathrooms, are by nature only intermittently occupied. When the required ventilation is mechanical, is it the intent of the code to require the system to be operating at all times the building is occupied to ensure ventilation whenever the room in question is used?

A: No. Because Section 401.3 only requires ventilation whenever a room or space is occupied, it would be permissible to activate the mechanical ventilation system serving that room by interconnection with the light switch in the room, a timer switch, interconnection with the building ventilation system, a separate manual switch in the room or another approved means.

SECTION 403 MECHANICAL VENTILATION

403.2.1 Recirculation of air. The air required by Section 403.3 shall not be recirculated. Air in excess of that required by Section 403.3 shall not be prohibited from being recirculated as a component of supply air to building spaces, except that:

1. Ventilation air shall not be recirculated from one dwelling to another or to dissimilar occupancies.
2. Supply air to a swimming pool and associated deck areas shall not be recirculated unless such air is dehumidified to maintain the relative humidity of the area at 60 percent or less. Air from this area shall not be recirculated to other spaces where 10 percent or more of the resulting supply airstream consists of air recirculated from these spaces.
3. Where mechanical exhaust is required by Note b in Table 403.3, recirculation of air from such spaces shall be prohibited. All air supplied to such spaces shall be exhausted, including any air in excess of that required by Table 403.3.
4. Where mechanical exhaust is required by Note h in Table 403.3, mechanical exhaust is required and recirculation is prohibited where 10 percent or more of the resulting supply airstream consists of air recirculated from these spaces.

Q: Is it the intent of Section 403.2.1 to prohibit the recirculation of air to a dissimilar occupancy, whether or not the dissimilar occupancy is classified in the same group or occupied by the same tenant?

A: Yes. Air quality is not a function of occupancy classification or tenancy, and the code provisions apply to buildings of single or mixed use, as well as buildings occupied by single or multiple tenants. Dissimilar occupancies will have different contaminants and/or odors with a varying nature and concentration.

The dissimilar occupancy concept is not limited to occupancies classified as different groups and could, therefore, apply to dissimilar occupancies within a single group. Likewise, dissimilar occupancies are not limited to multiple tenancy and may occur in a single tenant space. Recirculating air to a dissimilar occupancy from which it is withdrawn may subject the occupants thereof to contaminants and/or odors that are not inherent to the occupancy or workplace. Such contaminants and odors could be harmful to the occupants or judged objectionable.

Mixed use buildings may have different environmental air in each occupancy, and recirculation of air between such areas would be prohibited (e.g., offices, Group B; located in a factory, Group F). Conversely, mixed use buildings may have similar environmental air in each occupancy, and recirculation of air between such areas would be allowed (e.g., offices, Group B; located in a library, Group A-3). Buildings of a single occupancy classification may contain areas or incidental occupancies that have different environmental air, and recirculation of air between such areas would be prohibited (e.g., a conference room incidental to an office building with an occupancy in Group B or a locker/dressing room incidental to a theater with an occupancy in Group A-1).

403.3.2 Common ventilation system. Where spaces having different ventilation rate requirements are served by a common ventilation system, the ratio of outdoor air to total supply air for the system shall be determined based on the space having the largest outdoor air requirement or shall be determined in accordance with the following formula:

$$Y = \frac{X}{(1 + X - Z)} \qquad \textbf{(Equation 4-1)}$$

where:

$Y = V_{ot}/V_{st}$ = Corrected fraction of outdoor air in system supply.

$X = V_{on}/V_{st}$ = Uncorrected fraction of outdoor air in system supply.

$Z = V_{oc}/V_{sc}$ = Fraction of outdoor air in critical space. The critical space is that space with the greatest required fraction of outdoor air in the supply to this space.

V_{ot} = Corrected total outdoor airflow rate.

V_{st} = Total supply flow rate, i.e., the sum of all supply for all branches of the system.

V_{on} = Sum of outdoor airflow rates for all branches on system.

V_{oc} = Outdoor airflow rate required in critical spaces.

V_{sc} = Supply flow rate in critical space.

Q: Are the requirements in Section 403.3.2 only applicable when a variable air volume (VAV) air distribution system is used?

A: No. A VAV air distribution system is not required to be used for the provisions of Section 403.3.2 to be applicable. A fixed amount of supply air is permitted and can be supplied to the various spaces throughout the building.

CHAPTER 5

EXHAUST SYSTEMS

SECTION 507
COMMERCIAL KITCHEN HOODS

507.2 Where required. A Type I or Type II hood shall be installed at or above all commercial cooking appliances in accordance with Sections 507.2.1 and 507.2.2. Where any cooking appliance under a single hood requires a Type I hood, a Type I hood shall be installed. Where a Type II hood is required, a Type I or Type II hood shall be installed.

Q: The kitchen of a restaurant is provided with HVAC equipment sized to control the temperatures or moisture given off by various cooking equipment and appliances. Does Section 507.2 of the *International Mechanical Code* require commercial kitchen exhaust hoods for the cooking equipment and appliances?

A: Yes. Section 507.2 specifically requires Type I or Type II hoods at or above commercial appliances or dishwashers. As defined in Section 202, "Commercial cooking appliances" are those that are used for heating or cooking food. The commercial kitchen exhaust hood requirements are intended to abate not only grease-laden vapors, smoke and odors but also excessive amounts of heat and steam that put a burden on the HVAC system.

CHAPTER 6

DUCT SYSTEMS

SECTION 601 GENERAL

[B] 601.2 Air movement in egress elements. Corridors shall not serve as supply, return, exhaust, relief or ventilation air ducts.

Exceptions:

1. Use of a corridor as a source of makeup air for exhaust systems in rooms that open directly onto such corridors, including toilet rooms, bathrooms, dressing rooms, smoking lounges and janitor closets, shall be permitted, provided that each such corridor is directly supplied with outdoor air at a rate greater than the rate of makeup air taken from the corridor.
2. Where located within a dwelling unit, the use of corridors for conveying return air shall not be prohibited.
3. Where located within tenant spaces of 1,000 square feet (93 m^2) or less in area, utilization of corridors for conveying return air is permitted.

Q: Can a single HVAC duct system (supply, return, exhaust, relief, or ventilation) serve an exit access corridor and spaces open and/or adjacent to the exit access corridor where the air supplied into and returned out of the open and/or adjacent spaces is through the duct system and not through the exit access corridor?

A: Yes. Air supplied from a single HVAC duct system is permitted to serve an exit access corridor and spaces open and/or adjacent to the exit access corridor. Air supplied by the duct system into the spaces open and/or adjacent to the exit access corridor, either by the duct system or by other mechanical or nonmechanical means, is not permitted to be transferred into or through the exit access corridor as supply air, return air, make up air, ventilation or exhaust.

SECTION 602 PLENUMS

602.2.1 Materials exposed within plenums. Except as required by Sections 602.2.1.1 through 602.2.1.5, materials within plenums shall be noncombustible or shall have a flame spread index of not more than 25 and a smoke-developed index of not more than 50 when tested in accordance with ASTM E 84.

Exceptions:

1. Rigid and flexible ducts and connectors shall conform to Section 603.
2. Duct coverings, linings, tape and connectors shall conform to Sections 603 and 604.
3. This section shall not apply to materials exposed within plenums in one- and two-family dwellings.
4. This section shall not apply to smoke detectors.
5. Combustible materials enclosed in noncombustible raceways or enclosures, approved gypsum board assemblies or enclosed in materials listed and labeled for such application.

Q: Are exposed combustible thermal and acoustical materials such as ceiling tiles, batt and board insulation and piping insulation permitted to be located within a noncombustible plenum, thus forming part of the enclosure?

A: Yes. Section 602.2.1 of the *International Mechanical Code* specifically permits materials that are combustible to be exposed within a plenum provided that the materials have maximum flame spread and smoke-developed ratings of 25 and 50, respectively, when tested in accordance with ASTM E 84. These limitations recognize that these materials will be used on the interior surfaces of the plenum or will form a portion of the plenum itself and address the potential fire hazard.

Combustible suspended acoustical ceiling tiles, for example, are permitted to form the underside of a noncombustible plenum, provided that the material complies with the flame spread and smoke-developed limitations of Section 602.2.1.

SECTION 603 DUCT CONSTRUCTION AND INSTALLATION

603.8 Underground ducts. Ducts shall be approved for underground installation. Metallic ducts not having an approved protective coating shall be completely encased in a minimum of 2 inches (51 mm) of concrete.

Q: Is UL 181 the applicable standard for flexible nonmetallic underground ducts?

A: Yes. Sections 603.5 and 603.6 require nonmetallic and/or flexible ducts, respectively, to conform to the requirements of UL 181. Section 603.8 addresses the installation of underground ducts, and not the acceptability of duct materials and construction. However, compliance with UL 181 does not provide assurance that the ducts are rated for underground installation. Such suitability must be evaluated based on the manufacturer's installation instructions for the intended underground application and corresponding supportive evidence.

SECTION 606
SMOKE DETECTION SYSTEMS CONTROL

606.2.1 Return air systems. Smoke detectors shall be installed in return air systems with a design capacity greater than 2,000 cfm (0.9 m^3/s), in the return air duct or plenum upstream of any filters, exhaust air connections, outdoor air connections, or decontamination equipment and appliances.

> **Exception:** Smoke detectors are not required in the return air system where all portions of the building served by the air distribution system are protected by area smoke detectors connected to a fire alarm system in accordance with the *International Fire Code*. The area smoke detection system shall comply with Section 606.4.

Q: The attached drawing shows two air-conditioning units with a design capacity of 1100 cfm each with a separate return air system for each unit. The two air conditioning units serve a common area. Section 606.2.1 of the *International Mechanical Code* requires a smoke detector in return air systems with a design capacity greater than 2000 cfm.

Are smoke detectors required in the return air ducts or plenums where the supply air and the return air are ducted separately to each air conditioning unit? See Attachment No. 1.

A: No. Smoke detectors are not required in the return air ducts per the attached drawing. The capacity of the two air conditioning units is only added together when the units share a common supply duct, return air duct or plenum.

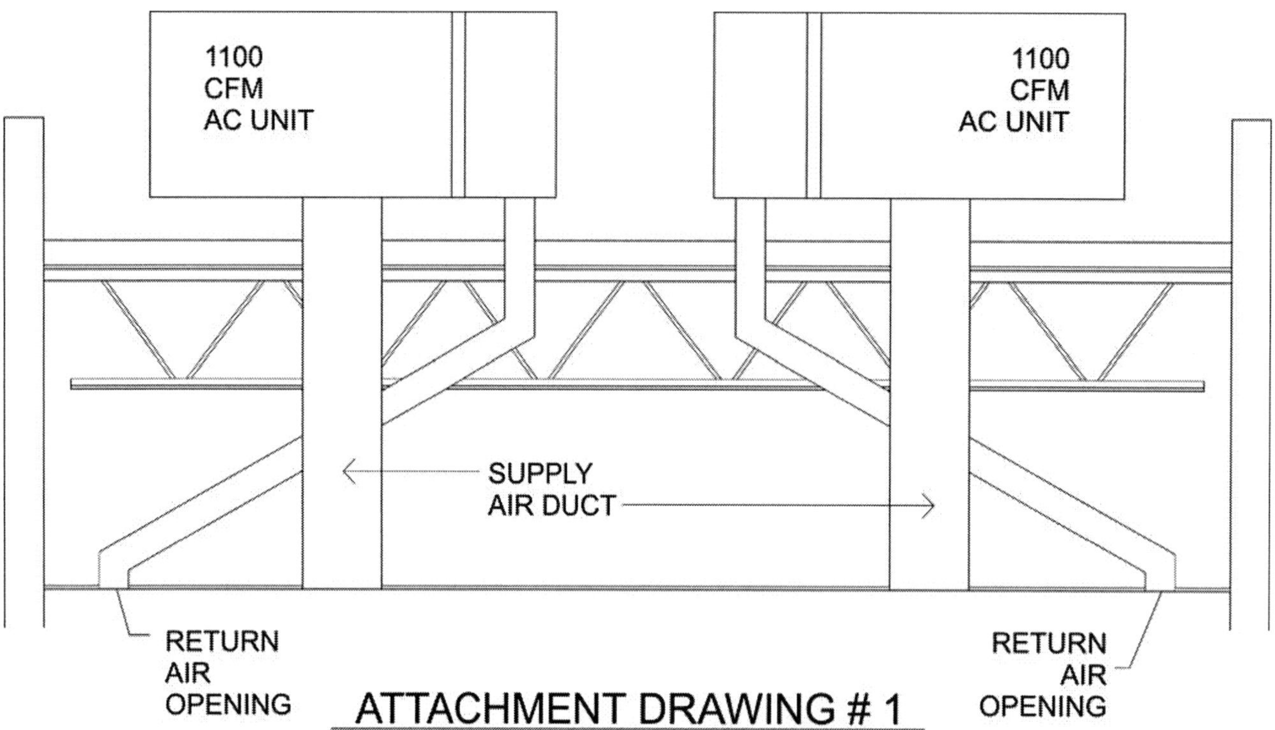

ATTACHMENT DRAWING # 1

CHAPTER 8

CHIMNEYS AND VENTS

SECTION 804
DIRECT-VENT, INTEGRAL VENT AND MECHANICAL DRAFT SYSTEMS

804.3.4 Horizontal terminations. Horizontal terminations shall comply with the following requirements:

1. Where located adjacent to walkways, the termination of mechanical draft systems shall be not less than 7 feet (2134 mm) above the level of the walkway.
2. Vents shall terminate at least 3 feet (914 mm) above any forced air inlet located within 10 feet (3048 mm).
3. The vent system shall terminate at least 4 feet (1219 mm) below, 4 feet (1219 mm) horizontally from or 1 foot (305 mm) above any door, window or gravity air inlet into the building.
4. The vent termination point shall not be located closer than 3 feet (914 mm) to an interior corner formed by two walls perpendicular to each other.
5. The vent termination shall not be mounted directly above or within 3 feet (914 mm) horizontally from an oil tank vent or gas meter.
6. The bottom of the vent termination shall be located at least 12 inches (305 mm) above finished grade.

Q: Are the provisions of Section 804.3.4, Item 3, applicable for determining the minimum separation distance between the termination of a horizontal vent equipped with a power exhauster and a gravity air inlet for combustion air?

A: Yes. The provisions of Section 804.3.4, Item 3, are intended to be used to establish separation distances between power exhauster discharges and air inlets. The requirements of Section 804.3.4 acknowledge the specific operating and performance characteristics of power-exhausted vents and, therefore, would govern in lieu of other relevant general mechanical code provisions.

804.3.5 Vertical terminations. Vertical terminations shall comply with the following requirements:

1. Where located adjacent to walkways, the termination of mechanical draft systems shall be not less than 7 feet (2134 mm) above the level of the walkway.
2. Vents shall terminate at least 3 feet (914 mm) above any forced air inlet located within 10 feet (3048 mm) horizontally.
3. Where the vent termination is located below an adjacent roof structure, the termination point shall be located at least 3 feet (914 mm) from such structure.
4. The vent shall terminate at least 4 feet (1219 mm) below, 4 feet (1219 mm) horizontally from, or 1 foot (305 mm) above any door, window or gravity air inlet for the building.
5. A vent cap shall be installed to prevent rain from entering the vent system.
6. The vent termination shall be located at least 3 feet (914 mm) horizontally from any portion of the roof structure.

Q: What constitutes a "walkway" as referenced in Item 1 of Sections 804.3.4 and 804.3.5?

A: The termination requirements for horizontal and vertical vents equipped with power exhausters are intended to protect individuals that may be walking near the vent's discharge from being exposed to the vent's products of combustion. These requirements are applicable to both public and private walkways. A walkway may be present even though a paved or other prepared walking surface does not exist. Any area that is normally expected to serve as a path of travel is considered a walkway for the purposes of protecting individuals from exposure to the vent's discharge.

Q: What is the minimum separation distance required between the termination of horizontal or vertical vents equipped with power exhausters and a "walkway" to not be considered adjacent to the "walkway"?

A: The intent of the code provisions is to protect individuals that may be walking near the vent's discharge from being exposed to the vent's products of combustion. The code does not explicitly prescribe a minimum distance suitable for all conditions and installations between a vent termination and a walkway. An evaluation of the worst-case conditions which result in the most adverse dispersion of exhaust discharge from the vent must be made. Several factors must be considered in establishing a minimum distance including, but not limited to, the location and direction of the vent discharge, the direction of prevailing winds, the volumetric flow rate and velocity of the vent's exhaust and the densities of the vent's exhaust and ambient air. As a matter not specifically provided for, the code official must evaluate these factors as well as the manufacturer's installation instructions for the labeled equipment and determine an appropriate distance to abate the hazards to an acceptable level.

CHAPTER 10

BOILERS, WATER HEATERS AND PRESSURE VESSELS

SECTION 1006
SAFETY AND PRESSURE RELIEF VALVES AND CONTROLS

1006.5 Installation. Safety or relief valves shall be installed directly into the safety or relief valve opening on the boiler or pressure vessel. Valves shall not be located on either side of a safety or relief valve connection. The relief valve shall discharge by gravity.

Q: Section 1006.5 requires the safety or relief valve to be installed directly into the opening on the boiler or pressure vessel. Most safety valves and relief valves have a female connection, and most boilers have a female connection for the safety or relief valve. What is the maximum length permitted for a pipe nipple to facilitate the installation of a safety or relief valve to be considered directly connected?

A: The code does not specify a maximum permitted length. However, the pipe nipple between the safety or pressure relief valve and the boiler or pressure vessel should not be any longer than what is necessary to make a proper installation. The maximum length that is usually necessary is 12 inches. The proper installation of the valve would require the nipple to pass through the outside jacket of the boiler. Furthermore, a safety or pressure relief valve must have the spindle installed in the vertical position. The distance between the valve and the boiler must be maintained at a minimum to prevent a blockage from developing in the stagnant length of piping. A stoppage in piping could prevent the valve from operating in a proper manner.

INTERNATIONAL CODE INTERPRETATIONS

FUEL GAS

2006

ICC
INTERNATIONAL CODE COUNCIL®

CHAPTER 3

GENERAL REGULATIONS

SECTION 305
INSTALLATION

305.3 Elevation of ignition source. Equipment and appliances having an ignition source shall be elevated such that the source of ignition is not less than 18 inches (457 mm) above the floor in hazardous locations and public garages, private garages, repair garages, motor fuel-dispensing facilities and parking garages. For the purpose of this section, rooms or spaces that are not part of the living space of a dwelling unit and that communicate directly with a private garage through openings shall be considered to be part of the private garage.

> **Exception:** Elevation of the ignition source is not required for appliances that are listed as flammable vapor ignition resistant.

Q: A private garage attached to a dwelling unit contains a 4-inch-high platform installed over a small portion of the garage floor. The required water heater for the residence will be located upon the platform. Does Section 305.3 of the *International Fuel Gas Code* and Section 304.3 of the *International Mechanical Code* require the 18-inch installation height for the ignition source to be measured from the top of the platform?

A: No. A private garage not only contains vehicles with fuel tanks but also is commonly used as a storage area for containers of gasoline and other hazardous household items such as paint, thinners and solvents. These flammable and combustible liquids produce vapors that are more dense than air. As these vapors collect on the garage floor, they represent a serous fire hazard due to the ignition potential of fuel-burning water heaters, mechanical equipment and appliances.

To reduce substantially the risk of explosion and fire, Section 305.3 of the *International Fuel Gas Code* and Section 304.3 of the *International Mechanical Code* require that fuel-burning water heaters, mechanical equipment and appliances be elevated such that their ignition source is not less than 18 inches above the garage floor. In this case, the 18-inch installation height can be measured from the main garage floor where the flammable vapors will accumulate. Note this assumes the raised platform is limited in size such that it will not be used for the storage of flammable and combustible liquids.

CHAPTER 4

GAS PIPING INSTALLATIONS

SECTION 411 APPLIANCE AND MANUFACTURED HOME CONNECTIONS

411.1 Connecting appliances. Except as required by Section 411.1.1, appliances shall be connected to the piping system by one of the following:

1. Rigid metallic pipe and fittings.
2. Corrugated stainless steel tubing (CSST) where installed in accordance with the manufacturer's instructions.
3. Semirigid metallic tubing and metallic fittings. Lengths shall not exceed 6 feet (1829 mm) and shall be located entirely in the same room as the appliance. Semirigid metallic tubing shall not enter a motor-operated appliance through an unprotected knockout opening.
4. Listed and labeled appliance connectors in compliance with ANSI Z21.24 and installed in accordance with the manufacturer's installation instructions and located entirely in the same room as the appliance.
5. Listed and labeled quick-disconnect devices used in conjunction with listed and labeled appliance connectors.
6. Listed and labeled convenience outlets used in conjunction with listed and labeled appliance connectors.
7. Listed and labeled appliance connectors complying with ANSI Z21.69 and listed for use with food service equipment having casters, or that is otherwise subject to movement for cleaning, and other large movable equipment.
8. Listed and labeled outdoor appliance connectors in compliance with ANSI Z21.75/CSA 6.27 and installed in accordance with the manufacturer's installation instructions.

Q: Item 4 of Section 411.1 permits gas appliances to be connected to a building's fuel-gas piping system via listed and labeled gas appliance connectors. Further, Section 411.1.2 requires that those connectors be installed such that they are protected from physical damage. Does this requirement limit the use of connectors to just those installations where the gas appliance itself provides the necessary protection against physical damage?

A: No. The use of connectors with their corrugated shape and flexible features is universal for gas-fired clothes dryers, ranges and ovens and is common for water heaters, furnaces, suspended radiant heaters and decorative appliances. These gas appliance connectors must be listed and labeled and are subject to stringent installation restrictions that are stated in the connector manufacturer's installation instructions. A primary concern with such connectors is that they must not be subjected to physical damage or repeated movement. Repeated bending, flexing and similar movement can work-harden the connector, making it brittle and subject to cracking. Care must be taken to protect the connector from damage caused by being impacted, bumped or moved by the building occupants, pets, side-swinging doors, machinery, vehicles, etc. Section 411.1.2 is not, however, intended to restrict gas appliance connectors to just out-of-sight locations behind or beneath appliances.

INTERNATIONAL CODE INTERPRETATIONS

PLUMBING

2006

INTERNATIONAL CODE COUNCIL®

CHAPTER 4

FIXTURES, FAUCETS AND FIXTURE FITTINGS

SECTION 403
MINIMUM PLUMBING FACILITIES

403.2 Separate facilities. Where plumbing fixtures are required, separate facilities shall be provided for each sex.

Exceptions:

1. Separate facilities shall not be required for dwelling units and sleeping units.
2. Separate facilities shall not be required in structures or tenant spaces with a total occupant load, including both employees and customers, of 15 or less.
3. Separate facilities shall not be required in mercantile occupancies in which the maximum occupant load is 50 or less.

Q: How is the number of employees determined for a mercantile occupancy?

A: For a mercantile occupancy, the design occupant load specified in the *International Building Code* is predominantly a customer occupant load. For the purpose of separate plumbing facilities, the number of employees would be the actual number as determined by the building owner or tenant based on the needs of the mercantile use, which is subject to the approval of the code official.

CHAPTER 6

WATER SUPPLY AND DISTRIBUTION

SECTION 608
PROTECTION OF POTABLE WATER SUPPLY

608.16.6 Connections subject to backpressure. Where a potable water connection is made to a nonpotable line, fixture, tank, vat, pump or other equipment subject to back-pressure, the potable water connection shall be protected by a reduced pressure principle backflow preventer.

Q: Does Section 608.16.6 require a reduced pressure principle backflow preventer to protect the potable water supply from an automatic sprinkler system without chemical additives or antifreeze when the sprinkler system is equipped with a fire department connection?

A: No. Although the potable water supply must be protected against backflow where it is connected to an automatic fire sprinkler system equipped with a fire department connection, Section 608.16.4 allows either a double check-valve backflow prevention assembly or a reduced pressure principle backflow preventer to be utilized to protect the potable water supply against potential contamination from the contents of the automatic sprinkler piping system. The intent of Section 608.16.6 is to provide minimum requirements for general applications not specifically addressed by the code.

Although a connection by the fire department during a fire will likely elevate the system pressure beyond the domestic water supply pressure (resulting in a "backpressure"), the general provisions of Section 608.16.6 are not intended to supersede the specific provisions of Section 608.16.4 that allow a double check-valve backflow prevention assembly as the minimum required backflow protection for automatic sprinkler systems.

A double check-valve backflow prevention assembly is tested and rated for backpressure applications typically associated with automatic sprinkler systems. This is consistent with the referenced standards indicated in Section 608.13.7. In addition, Table 608.1 states that double check-valve assemblies are limited to low-hazard backpressure or backsiphonage applications. The purpose of a double check-valve backflow preventer is to keep polluted water from flowing into a potable water distribution system when the pressure of the polluted part of the system is temporarily higher than the potable water piping.

It should be noted that a double check-valve assembly is only rated for low-hazard applications where the potential contamination source does not create a threat to public health but is capable of adversely and unreasonably affecting the aesthetic qualities of the potable water supply. High-hazard applications where the automatic sprinkler systems contain chemical additives, antifreeze, or that can be served by a nonpotable external water source that is a hazard to public health through poisoning or the spread of disease is required to be isolated from the potable water supply by a reduced pressure principal backflow preventer. The code official and system designer must investigate all potential secondary water sources used for fire fighting, which can be introduced into the fire department connection, to establish the degree of hazard before determining the appropriate type of backflow prevention device.

CHAPTER 9

VENTS

SECTION 903
OUTDOOR VENT EXTENSION

903.2 Vent stack required. A vent stack shall be required for every drainage stack that has five branch intervals or more.

Q: Does a waste stack vent system, as described in Section 910, that is 5 branch intervals or more in height require a vent stack?

A: Yes. Section 903.2 literally says that a vent stack is required for all drainage stacks that are 5 branch intervals or more, and this would not exclude waste stack vents.

The intent of Section 910 is that the waste stack vent is in itself a venting method. A waste stack vent is oversized compared to other drainage stacks. The technical issue is whether or not a vent stack is needed to relieve positive pressure at the base of a waste stack vent system greater than 5 branch intervals in height. In this situation, such pressure fluctuations are likely to be negligible and, therefore, the need for a vent stack is questionable; however, one cannot ignore the literal wording of Section 903.2 and its lack of an exception for waste stack vents.

CHAPTER 10

TRAPS, INTERCEPTORS AND SEPARATORS

SECTION 1003 INTERCEPTORS AND SEPARATORS

1003.4 Oil separators required. At repair garages, car-washing facilities, at factories where oily and flammable liquid wastes are produced and in hydraulic elevator pits, separators shall be installed into which all oil-bearing, grease-bearing or flammable wastes shall be discharged before emptying into the building drainage system or other point of disposal.

Exception: An oil separator is not required in hydraulic elevator pits where an approved alarm system is installed.

1003.4.2.2 Garages and service stations. Where automobiles are serviced, greased, repaired or washed or where gasoline is dispensed, oil separators shall have a minimum capacity of 6 cubic feet (0.168 m^3) for the first 100 square feet (9.3 m^2) of area to be drained, plus 1 cubic foot (0.28 m^3) for each additional 100 square feet (9.3 m^2) of area to be drained into the separator. Parking garages in which servicing, repairing or washing is not conducted, and in which gasoline is not dispensed, shall not require a separator. Areas of commercial garages utilized only for storage of automobiles are not required to be drained through a separator.

Q: The waste effluent from a car wash passes through a sand interceptor before discharging to the sanitary drainage system. Is an oil interceptor required to be installed in addition to a sand interceptor for automatic and manual car-washing facilities?

A: No. The mere presence of an automatic or manual car wash does not in itself necessitate the installation of an oil separator. A sand interceptor is capable of accommodating the effluent from a car wash. The material present in the effluent from a car wash includes mud, sand, sediment and sludge, in addition to small amounts of greasy and oily waste.

The intent of Section 1003 is to collect and safely store substances and materials that are detrimental to the drainage system, sewage treatment facility, or private sewage disposal system. Therefore, a case-by-case analysis of each vehicle-washing facility, considering the anticipated conditions of use, must be performed prior to the selection of the appropriate interceptor/separator.

Where the anticipated volume of oil and greasy waste exceeds the retention capabilities of the sand interceptor, or where local regulatory agencies or sewage treatment facilities impose oil and grease discharge limitations, an oil separator is required.

The first sentence of Section 1003.4.2.2 is not intended to mandate that wash water from car washes pass through an oil separator. The requirement is intended to provide minimum design criteria for oil separators serving motor vehicle garages or service stations where, among other service-related activities, machinery, vehicles or parts thereof are cleaned in preparation for repairs.

Q: A separator is provided for a garage/service station in which automobiles are repaired and serviced. The garage bay has a floor area of 150 square feet. Does Section 1003.4.2.2 of the *International Plumbing Code* require the separator to have a minimum capacity of 6.5 cubic feet?

A: Yes. The intent of the code is to require a separator for those floor areas that are used for lubricating, changing oil, dispensing fuel or other repair work of motor vehicles. Where floor drains are provided in a garage or service station, flammable and combustible liquids may enter the drainage system unless a separator is provided. A separator is required to be sized based on the floor area that drains into it. For floor areas greater than 100 square feet in area, the size of the separator is based on a linear relationship of Capacity = 5 + Area/100. In the case of a 150-square-foot garage bay, the separator must have a minimum capacity of 6.5 cubic feet. It is not necessary to round the capacity of the separator up to the next whole number.

INTERNATIONAL CODE INTERPRETATIONS

EXISTING BUILDING

2006

ICC
INTERNATIONAL CODE COUNCIL®

CHAPTER 7

ALTERATIONS - LEVEL 2

SECTION 704 FIRE PROTECTION

704.2.2 Groups A, E, F-1, H, I, M, R-1, R-2, R-4, S-1, and S-2. In buildings with occupancies in Groups A, E, F-1, H, I, M, R-1, R-2, R-4, S-1, and S-2, work areas that include exits or corridors shared by more than one tenant or that serve an occupant load greater than 30 shall be provided with automatic sprinkler protection where all of the following conditions occur:

1. The work area is required to be provided with automatic sprinkler protection in accordance with the *International Building Code* as applicable to new construction;
2. The work area exceeds 50 percent of the floor area; and
3. The building has sufficient municipal water supply for design of a fire sprinkler system available to the floor without installation of a new fire pump.

Exception: Work areas in Group R occupancies three stories or less in height.

GIVEN: Assume Level 2 alterations to an existing building that is not classified as a high-rise building. The building is not a Group R occupancy. All the listed conditions, including Conditions 1 and 2, are applicable. The municipal water supply for the building is sufficient for design of a fire sprinkler system to the floor containing the work area, without requiring the installation of a new fire pump.

Q: Is a fire sprinkler system required to be provided to the work area if a new water service pipe is required to be installed between the municipal water supply and the building?

A: No. Section 704.2.2, Condition 3 indicates that "The building has sufficient municipal water supply for design of a fire sprinkler system." Therefore, it was not intended that new water service pipes be installed from the water main to the building.

Q: Is a fire sprinkler system required to be provided to the work area if a new water distribution pipe (or riser) is required to be installed between the water service pipe and the work area?

A: No. Sprinkler system will not be required if a new riser must be constructed to bring water from lower floors.

STAFF COMMENTARY: For both of these answers, please note that the language in Section 704.2.2 is distinctly different from that found in Section 804.1.1 (Level 3 Alterations, high-rise buildings), where sprinklers will be required if the municipal water main at the site has sufficient municipal water supply.

One of the main resource documents upon which the *International Existing Building Code* was based is the Nationally Applicable Recommended Rehabilitation Provisions (NARRP). The origin of many of the *International Existing Building Code* sections can be traced to similar provisions in the NARRP. The NARRP does not require sprinkler systems in work areas for the level of construction called "Alterations" (this is approximately the same level as *International Existing Building Code* Level 2 Alterations and is found in NARRP Chapter 5). Section 606.1 of NARRP for the level called "Reconstruction" (approximately similar to Level 3 Alterations of *International Existing Building Code*), requires sprinkler system, but provides an exception if "...an automatic water supply for sprinkler protection is not available at that floor, the building official shall be permitted to accept alternative protection." A review of this history indicates the intent of the *International Existing Building Code* drafting committee to require sprinkler systems in Level 2 Alterations only if sufficient municipal water supply is available at the floor under discussion.

INTERNATIONAL CODE INTERPRETATIONS

ENERGY CONSERVATION

2006

ICC
INTERNATIONAL CODE COUNCIL®

CHAPTER 4

RESIDENTIAL ENERGY EFFICIENCY

SECTION 402
BUILDING THERMAL ENVELOPE

402.5 Moisture control. (Mandatory). The building design shall not create conditions of accelerated deterioration from moisture condensation. Above-grade frame walls, floors and ceilings not ventilated to allow moisture to escape shall be provided with an approved vapor retarder. The vapor retarder shall be installed on the warm-in-winter side of the thermal insulation.

Exceptions:

1. In construction where moisture or its freezing will not damage the materials.
2. Frame walls, floors and ceilings in jurisdictions in Zones 1 through 4. (Crawl space floor vapor retarders are not exempted.)
3. Where other approved means to avoid condensation are provided.

Q: The requirements in Section 402.5 stipulate that a vapor retarder must be installed on the warm-in-winter side of the thermal insulation. Is the installation of the vapor retarder required continuously across the face of the wall studs?

A: No. The word continuous is not used in the referenced code section. Therefore, a vapor retarder is not required to be installed in a manner that will provide a continuous barrier on the warm-in-winter side of the exterior wall.